U-1105 'BLACK PANTHER'

DEDICATION

For US Navy Capt Hubert "Hugh" Murphy (1917–2013) . . .
. . . and the US and Royal Navy prize crew of the German *U-1105* "Black Panther".
They braved the North Atlantic winter storms of 1945–46 to bring this unique U-boat to the USA without ever knowing the reason why.

GERMAN SUBMARINE
U-1105
'BLACK PANTHER'
THE NAVAL ARCHAEOLOGY OF A U-BOAT

AARON STEPHAN HAMILTON

OSPREY PUBLISHING
Bloomsbury Publishing Plc
PO Box 883, Oxford, OX1 9PL, UK
1385 Broadway, 5th Floor, New York, NY 10018, USA
E-mail: info@ospreypublishing.com
www.ospreypublishing.com

OSPREY is a trademark of Osprey Publishing Ltd

First published in Great Britain in 2019

© Text and photographs Aaron Stephan Hamilton, 2019

Aaron Stephan Hamilton has asserted his right under the Copyright, Designs and Patents Act, 1988, to be identified as Author of this work.

All rights reserved. No part of this publication may be reproduced or transmitted in any form or by any means, electronic or mechanical, including photocopying, recording, or any information storage or retrieval system, without prior permission in writing from the publishers.

A catalogue record for this book is available from the British Library.

ISBN: HB 9781472835819; ePub 9781472835802; ePDF 9781472835826; XML 9781472835833

19 20 21 22 23   10 9 8 7 6 5 4 3 2 1

Printed in China through World Print Ltd.

Front cover: *U-1105* possibly preparing for one of its numerous tests in Loch Goil while commissioned with the Royal Navy (Royal Navy Submarine Museum).
Back cover: Top: *U-1105*'s Alberich coating along the portside bow – for more detail see p.29 (Maryland Historical Trust).
Top centre: *U-1105* on the day it was commissioned, June 3, 1944 (Maryland Historical Trust).
Below centre: Base of the sky scope – for more detail see p.116 (Paul D. Lenharr II).
Bottom: 2cm watertight ammunition canister located aft of the conning tower on the upper Wintergarten (gun deck) (Paul D. Lenharr II).

Osprey Publishing supports the Woodland Trust, the UK's leading woodland conservation charity.

To find out more about our authors and books visit www.ospreypublishing.com. Here you will find extracts, author interviews, details of forthcoming events and the option to sign up for our newsletter.

Figure 1. "Black Panther".
(Courtesy of Maryland Historical Trust)

This image is the original representation of the *U-1105*'s unique conning tower emblem that was handmade by a former crewman. The origin of the emblem stems from three events. First, the fact that members of the crew enjoyed a lively jazz song titled "Black Panther" that was obtained on record and played on board during its training while in the Baltic. The U-boat's captain, however, was no fan of jazz. He preferred opera and banned the playing of the song on board. Second, the captain's last name was Schwarz, which is the German word for "black". Third, the U-boat was black due to its Alberich coating, while all other German U-boats were a variation of gray. When the time came for an emblem to be designed, the crew raised the prospect of a "Schwarz (Black) Panther" sitting on top of a globe portraying the U-boat's operational area of the North Atlantic. The captain, noting the play on both words and themes, agreed. As one former crewman recounted: "So we had a name, an emblem, and even a theme song!"

CONTENTS

Foreword by Dr Innes McCartney — 6
Preface — 8
Acknowledgments — 10
Introduction — 12

Chapter 1	Late-war Technical and Tactical Evolution	16
Chapter 2	Construction and Training	34
Chapter 3	First and Last Wartime Patrol	46
Chapter 4	Postwar Evaluation and Testing by the Royal Navy	62
Chapter 5	North Atlantic Transit to the USA	80
Chapter 6	Salvage Training and Depth Charge Testing in the Potomac River	90
Chapter 7	Forgotten but Never Lost	102
Chapter 8	Diving the *U-1105* Today	106
Chapter 9	Archeology of a Late War U-boat	109
Appendix A	Technical Specifications	128
Appendix B	The Crew	128
Appendix C	Chronological History of *U-1105*	130

Endnotes — 132
Index — 135

FOREWORD

Many years ago I had the opportunity to dive on some of the U-boat wrecks on the East Coast of the USA, including *U869* and *U853*. Although I didn't manage to visit *U1105*, I read about its fascinating rubber coating; code named "Alberich", it was one a number of truly transformational submarine technologies the German Navy developed in the mid-20th century, although little was publicly known then of its detailed characteristics.

Coincidentally, I had no idea that upon my return home to the UK I would almost immediately find myself looking at a previously unnoticed Alberich-coated U-boat in the waters of the English Channel. Alberich was the key to our successful identification of this U-boat wreck as *U480*, hundreds of miles from where it was listed as being sunk in 1945. The importance and rarity of its rubber coating was only known to me because of the rediscovery of *U1105* a few years beforehand.

In those days, what we actually knew about the technical features of the late-WWII U-boat, with its plethora of new technologies, such as Alberich, GHG Balkon, and *Snorchel*, was not great. It has taken a lot of dedicated archival and archaeological research to develop a better understanding of these and many other aspects of the late-war U-boats and their operations. However, there still remains much room for improvement.

Aaron Hamilton's study of *U1105*, the "Black Panther", is a very fine example of the new type of research that has helped to reshape our collective understanding of U-boat warfare and technology. He can do this because not only is he an assiduous historical researcher with a proven track record, but also because he is a diver. I have found that the ability to dive on submarine wrecks brings a practical "hands-on" experience that is hard to beat. It also can provide an inspiring spark that, once ignited, can drive truly great research. This book is a fine example of such a journey.

Within these pages is a great deal of information that was new to me, especially relating to the wartime performance of *U1105*. What's more, her life in the postwar period is truly fascinating; the voyage to the USA is a feat of seafaring never repeated. Mr Hamilton has also done much to show how important *U1105* was as a test subject. All in all, this is a fine study of a shipwreck, combining a descriptive approach to the wreck with detailed historical background to produce an excellent and informative book on the U-boat as a vessel of war and as an important artifact of our collective cultural heritage. It should serve as a benchmark for future studies of this type.

Dr Innes McCartney
Naval Archaeologist
Bournemouth University, UK, February 2018

PREFACE

My interest in the German U-boat that rests in the mud off Piney Point, Maryland was driven by a desire to better understand the late-war German technology with which it was outfitted, and how this technology was employed when it conducted its first and only wartime patrol. When maritime archeologists conducted the first survey on *U-1105* in the early 1990s, their understanding of this U-boat's unique features and wartime operational impact was based upon the available published historic record of that time, as well as what could be derived from the Naval Historical Center (now Naval History and Heritage Command) in Washington, DC, and the private German veterans institution known as the U-boat Archive in Cuxhaven, Germany. All of this research, however, was limited. The historic picture that was produced of *U-1105* at that time was certainly fascinating, but left many questions unanswered. Key features on the U-boat went undocumented. Aspects of its wartime history that appeared conflicted were not resolved. In the introduction to *U-1105*'s 1993 site survey for the Maryland Historical Society the main authors and lead maritime archeologists for the project, Michael Pohuski and Donald Shomette, wrote that: "Obviously additional research is necessary…" Indeed it was.

Like all shipwreck sites, measurements can be made, photographs taken, and drawings rendered. Yet, no matter how many dives are made, the shipwreck's complete history can never be divined through the interpretation of what rests on the sea floor, or in this case, the river bed. Maritime archeology can be used to answer lingering questions of history, and historical research can provide answers to questions raised during such surveys. However, maritime archeology and thorough archival research are forever critically linked. In this case, neither discipline alone could provide all the answers necessary to complete the history of *U-1105*.

My first dive on the *U-1105*, also known as the "Black Panther", set in motion a nearly six-year research effort that spanned two continents and nearly a dozen archives, libraries, and private collections. That effort resulted in my forthcoming study *Total Undersea War: The Evolutionary Role of the Snorkel in Dönitz's U-Boat Fleet, 1944-1945*. What I came to learn in the course of this research is that *U-1105* is not just unique, but one of a kind: it was the only U-boat ever to conduct a wartime patrol equipped with three distinct and transformative late-war technologies that ushered in the evolution of modern submarine warfare. This fact has gone unrecognized to this day.

It was clear to me after my very first dive on *U-1105* that this U-boat required its own extensive historical treatment given its unique place in the history of World War II, postwar testing, and the local history of Chesapeake Bay. As my dive partner Fred Engle likes to say: "It's a U-boat in our own backyard." As true as this is, unfortunately, no individual work of history has been published on *U-1105* to date. The few published articles and online references *to U-1105* are incomplete or inaccurate. May this work serve as a comprehensive historical and maritime archeological guide to one of the most unique submarines of World War II.

Aaron S. Hamilton
Fairfax, Virginia
May 2018

ACKNOWLEDGMENTS

Many individuals from around the world contributed in no small measure to this work.

Here in the USA, Dr Susan Langley and Troy J. Nowak provided access to the Maryland Historical Trust's archival record of *U-1105* that initiated my research back in 2013. Both Donald G. Shomette and Michael Pohuski shared their personal experiences of diving *U-1105* and the survey work conducted in the 1990s. David Howe offered insight about the wreck's local lore few know. Fred Engle, a former US submariner, served as my dive partner on several trips to the *U-1105* and provided me with access to all his film footage and graciously read through various versions of the manuscript. Many of the excellent underwater stills in this book were taken from video footage shot in 2014 by Paul D. Lenharr II, who owns and operates the Southern Maryland Divers, LLC.

No one has dived upon *U-1105*'s conning tower more than Tom Edwards and he was featured in several local documentaries about the dive site. He knows every square inch of the wreck above the mud line. His enthusiasm for *U-1105* is contagious. Over the course of decades, Tom has documented the wreck site through his video logs and shared those with me. Depending on the year and the conditions, aspects of the wreck appear that few have ever seen, such as the top of the "N.16" pennant number painted on its side when in it was in Royal Navy service after the war. He turned diving the "Black Panther" into a science, and I count myself lucky to have dived the site with him.

Janet Murphy, the daughter of US Navy Capt Hubert "Hugh" Murphy (1917–2013), who sailed *U-1105* from England to the USA in the winter of 1945–46, provided documents, photos, and most important, insight into her father's experiences with this German U-boat. Indeed, she revealed that for this US Navy submarine veteran of World War II, sailing *U-1105* across the North Atlantic proved the highlight of his military career.

From Canada, US Navy (Rtd) Capt Jerry Mason, who operates arguably the most authoritative online U-boat research website, www.uboatarchive.net, supported my work with enthusiasm from the very start. He provided access to the documents and U-boat war diaries necessary to build the complete picture of *U-1105*'s operational environment at the end of the war.

In England, Duncan Rogers helped arrange for historian Michael LoCicero to visit The National Archives and researcher Ed Gosling to visit the Royal Navy Submarine Archives and acquire specific archival records. George Malcolmson and Debbie Corner from the Royal Navy Submarine Museum Archives in Gosport identified a trove of *U-1105* documents and photos critical to this study.

Chief among all those who contributed to this project is Royal Air Force Air Cdre Derek Waller (Rtd) who provided without hesitation all of his research on *U-1105*'s postwar history acquired over years of dogged archival investigation. His knowledge of postwar U-boat trials and testing is unsurpassed and he is always willing to answer questions and offer guidance that I highly value.

In Germany, my 2017 trip to the U-boot Archiv in Altenbruch was extremely productive and served to finalize the research for this book. This trip's success was due to the gracious support of Frau Annemarie Bredow, Peter Schulz, and especially Rainer Stührenberg, who located a hoard of relevant snorkel documents that I required. Former Bundesmarine Officer Peter Monte was always supportive and responded to all my e-mail questions to the U-boot Archiv through the years. Kai Steenbuck, an avid U-boat historian and volunteer at the U-Bot Archiv, provided key technical observations of the snorkel exhaust system that proved of great value to my research.

Thank you all for supporting this 20-plus-year journey!

INTRODUCTION

I first became aware of *U-1105* in 1995 while a graduate student of military history at Old Dominion University. At the time I was working on a Master's degree focused on how the German armed forces in Western Europe managed to recover after their disastrous defeat in France during the summer of 1944 and continue to resist for nearly 12 more months. While my research did not focus on aspects of the German Navy, I knew anecdotally that U-boats operated in European coastal waters with ferocity long after they were driven from the North Atlantic.

It was reported in the news at that time of the discovery of a "lost" U-boat known as the "Black Panther" had occurred. Its unusual name was derived from a late-war secret rubber coating designed to reduce its sonar signature underwater. Its historic allure only increased as news reports cited its seemingly mysterious disposal by the US Navy in the Potomac River. Being a certified scuba diver, I longed to "get wet" and see *U-1105* for myself, going so far as to call the then director of the Maryland Historical Trust and enjoying a lengthy conversation about *U-1105* and how best to dive the site. However, it took me almost 20 years to finally do it and touch down upon its silted conning tower off Piney Point, Maryland.

My first dive on *U-1105* was extremely memorable. It took place on a Saturday in October 2013 right after a late-season storm had blown through. My dive partner was Tom Edwards. No person has dived *U-1105* more than Tom Edwards. He knows every inch of the dive site and served as a volunteer who placed and recovered the *U-1105* floating marker buoy for years for the Maryland Historical Trust. We left the long boat dock in a nearly 60-year-old converted Chesapeake crab boat in the dark, as the storm's lingering wind whipped the diminishing rain against us and our gear.

As we motored out to the dive site around Piney Point, Tom gave me the obligatory safety brief. The dive site would be dark, almost as black as night. If you lost sight of the U-boat you might find yourself disoriented in the inky blackness of the Potomac River. If you did, you would have to launch a surface marker buoy and ascend to the surface on your own, which could be dangerous as this was a high-traffic area of the river. If you surfaced on your own, you might have to drift with the current until the dive boat came and recovered you. The best was yet to come. Tom produced two pictures of bull sharks. Yes, bull sharks. One was caught in the Potomac River just above the *U-1105* dive site and the other just below. They were among the largest bull sharks caught in the river to date. Bull sharks, being able to live in salt and fresh water, are common visitors to the Chesapeake Bay in the late summer and early fall. I'll never forget Tom's words: "They'll see you before you see them."

By the time we made it out to the site we had noted that the dive ball that allowed you to descend to the conning tower was gone – possibly blown away by the recent storm. Tom decided he was going to dive to the bottom of the Maritime Preserve's main marker buoy chain then use a compass to make his way to the U-boat and send up a surface marker buoy tied to the wreck that I could then use as a dive guide. He rolled over the side of the boat and into the murky green swells. The rain had stopped by now, but the wind was still gusty underneath the gray sky.

Perhaps ten minutes later he surfaced off the starboard side of the dive boat and began to swim back to us running a thin orange line. When he made it back to the boat he informed me that the original dive ball was there, just encrusted enough with marine growth that the added weight forced it below the surface.

He had tied a line to it and told me to jump in, follow the line to the dive ball, and wait for him at 6m (20ft) below the surface, where he would join me shortly. Once he arrived on the dive line, we would descend to the conning tower of *U-1105* together.

I followed his instructions and soon found myself at what can only be described as a stark transition between a deep murky green and inky blackness. Even the line that held the dive ball in place showed no growth below this point, as the lack of light prevented the required photosynthesis. There I waited, staring at the line below as it disappeared into blackness, gently bobbing with the swells.

Minutes ticked by as my air supply diminished. I looked up the line and saw no Tom. I looked down the line again into the inky blackness. I knew I should have stuck around for Tom, but I had waited a long time for this dive. I made the decision to descend down the line on my own and rendezvous with Tom below.

Hand over hand I pulled myself down, armed with only a single canister light mounted to my wrist, its beam barely penetrating the blackness. After a few minutes that seemed much longer, I checked my depth gauge; I was at 18m (60ft) and I knew I was getting close. Then I felt the bump along my left foot. I froze, Tom's words echoing in my ears: "They'll see you before you see them." I slowly moved my left foot in the hope that I would again bump something stationary, because if I didn't, I knew instinctively I was not alone. Then I felt it.

My foot had struck cold steel of the port-side conning tower. I breathed a sigh of relief and slowly began to rotate my light around through the blackness as *U-1105* finally began to emerge from my imagination into reality. Although the majority of the U-boat is unfortunately submerged under river mud and inaccessible, you can still swim along the conning tower and see its rubber coating and the external snorkel trunking.

The *U-1105* "Black Panther", a Type VIIC German U-boat, is the only diveable U-boat in the world equipped with three unique late-war German technologies designed to transform what was essentially a submersible at the start of the war into a true submarine and enter a "Total Undersea War", as it was called by U-boat command. The most critical technology that *U-1105* was equipped with was an airmast device known as the snorkel that allowed a U-boat to obtain fresh air while remaining submerged. Subordinate to that new innovation was an antiacoustic coating known as "Alberich" that reduced the effectives of Allied sonar, and a new passive sonar array design called the GHG Balkon that gave a submerged U-boat a significant advantage in detecting Allied vessels. In fact, *U-1105* was the only U-boat in the entire fleet to conduct an operational patrol while equipped with all three late-war innovations.

After the war, *U-1105* was tested extensively by the Royal Navy; in fact, it was tested more so than any other surrendered U-boat. While most might believe its antiacoustic coating was at the heart of this testing, they would be wrong to draw that conclusion. Its snorkel took center stage. The main testing against Alberich occurred in a lab to confirm whether or not it had antiradar or antiacoustic properties. Alberich's operational relevance was slow to be implemented among Western Allied navies in the postwar period. Not even the US Navy conducted a single operational evaluation of any of *U-1105*'s late-war technology. They simply treated it as a derelict test subject.

What follows is this U-boat's fascinating story of its commissioned wartime service in the German Navy; its extensive testing as a Royal Navy submarine; its brief focus as a war prize sought after by the Soviets; its harrowing crossing of the North Atlantic – the last wartime U-boat to make such a crossing under its own power; and final allocation to the US Navy, who sunk it ingloriously during an explosives test, sending it to the river bottom for the sixth and final time. All aspects of *U-1105* visible above the mud line of the Potomac River where it rests today are thoroughly detailed and documented. Any diver who descends into the darkness of the Potomac River to see *U-1105* for themselves will now have a clear understanding of its features and the role they played in evolving the U-boat force in World War II.

One final point must be made: *U-1105* may have been forgotten through the decades after its final sinking in the Potomac River, but it was never "lost".

CHAPTER 1
LATE-WAR TECHNICAL AND TACTICAL EVOLUTION

At the start of World War II German U-boats spent the majority of their time operating surfaced. They developed wolfpack tactics designed to conduct coordinated attacks on Allied convoys at night that required a near-constant stream of spotter reports and updates back to U-boat central operational control, known as Befehlshaber der U-Boote (BdU). BdU then redirected the combat action to meet the changing tactical situation of the convoy. By early 1943, wolfpack attacks against Allied convoys proved a strategic threat to Britain's survival and the subsequent build-up of forces for a future invasion of Europe. The U-boat had to be defeated.

The main weakness of a U-boat was that it was a submersible and not a true submarine capable of remaining submerged indefinitely. U-boats had to spend a significant amount of time on the surface to maximize their speed in order to maneuver toward an enemy or transit danger areas quickly while running their diesel engines, as movement underwater using

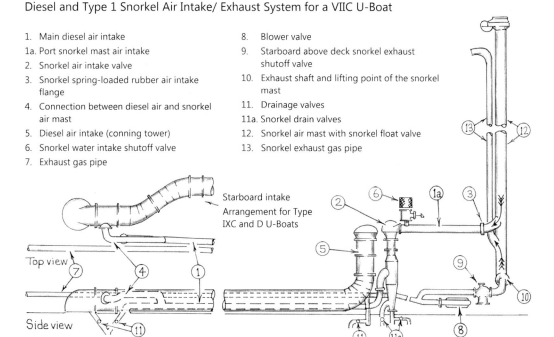

Figure 2. Type I flange snorkel operation (port side intake and starboard side exhaust).

batteries was slow. The entire hull design of a U-boat was conceived to maximize surface, not submerged cruising. When attacked on the surface a U-boat could dive but could only spend about 30 hours submerged before they had to refresh their oxygen supply. Allied technology and tactics evolved to meet this submersible threat.

With the cracking of the German Navy's Enigma code as a result of Operation *Ultra*, most encrypted messages sent between U-boat command and U-boats were deciphered and read, giving the Allies a good indication of a U-boat's location and bearing. The introduction of radar on surface ships and aircraft significantly improved their ability to locate a surfaced U-boat at night. Allied escort carriers with radar-equipped aircraft armed with depth charges and acoustic homing torpedoes now gave the Allies a marked advantage when it came to finding and sinking a U-boat anywhere in the mid-Atlantic. In what became known as "Black May" 1943, more U-boats were sunk in a single month than at any other point in the war, causing Adm Karl Dönitz to withdraw his U-boats temporarily from the

North Atlantic and cease anticonvoy operations. New technology and tactics had to be pioneered in order for his U-boats to survive and be able to regain the advantage against convoys operating in the Atlantic.

Dönitz tried numerous technical solutions over the course of the next 12 months in order to counter Allied radar and aircraft, with most initiatives failing to achieve what he hoped. Besides the North Atlantic, the Allies had focused in on the U-boat's main choke point, the Bay of Biscay. In order for U-boats to deploy into the Atlantic they had to make it from the French bases across this body of water. This journey could take a couple of days' surfaced cruising on diesels motors and expose U-boats to attack by Allied radar-equipped aircraft. If they tried the passage submerged on batteries their transit time nearly doubled and a U-boat still had to surface to recharge its batteries and refresh its oxygen. The solution proposed by Dönitz's principal U-boat engineer and innovator, Dr Hellmuth Walter, was an airmast termed the "*Schnorchel*" in German or "snorkel" in English.

THE SNORKEL

The snorkel was proposed in early 1943 as a means to solve the problem of having to spend time on the surface transiting the Bay of Biscay from French U-boat bases to the Atlantic Ocean and consequently be exposed to detection and possible destruction by Allied radar-equipped aircraft. It became the cornerstone of what Walter termed "Total Undersea War", which meant evolving all aspects of the surface-based U-boat into those of a true submarine that no longer had to surface. It took about six months for the first operational prototype to be fielded and about ten months for the first retrofits of the diesel U-boat fleet to begin.

No other Kriegsmarine technical development during World War II had a greater impact on U-boat operations than the introduction of the snorkel. It was the snorkel that introduced the evolutionary next step in submarine warfare that transformed submersibles into submarines. The snorkel was the first piece of German U-boat technology adopted by all navies around the world after the war. Indeed, today, the snorkel remains a prominent feature on even modern nuclear-powered submarines.

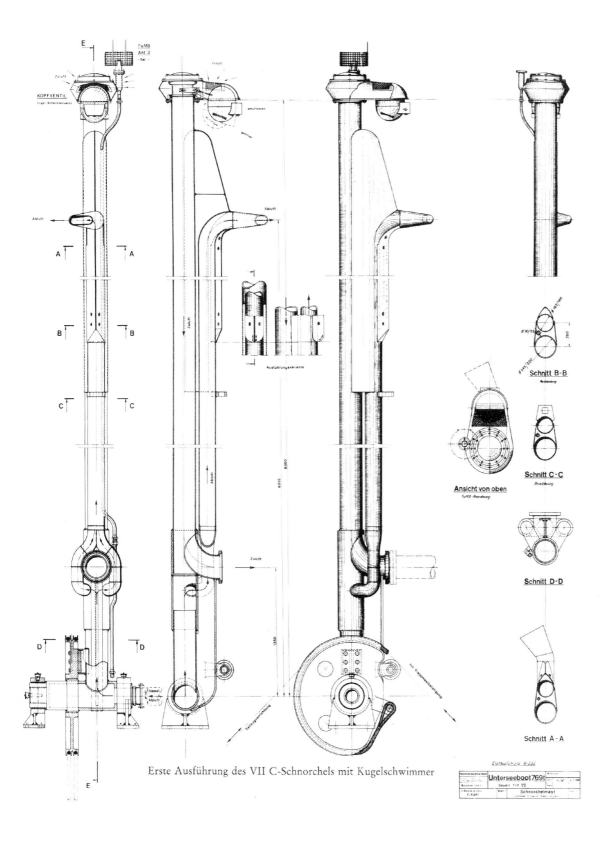

Figure 3. Diagram of the Type I flange snorkel mast. The intake and exhaust trunking were not encased within a single tube in the Type I flange version. Instead, the exhaust trunk bifurcated as it exited the mast at the base of the flange, then rejoined into one exhaust pipe at the top of the flange before ending at the port outlet approximately 1m (3ft 3in.) below the ball float. A subsequent modification, evident on the U-1105, encased both intake and exhaust lines into a single streamlined mast, though the exhaust line still bifurcated around the flange. Also note that this early version depicts a pulley lift system, while the U-1105 had the later piston version. (Author's collection)

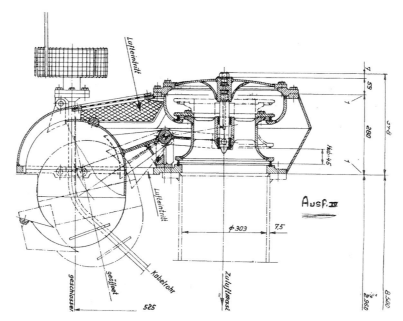

Figure 4. Ball float head valve with radar detection dipole aerial. The design was simple and effective: gravity kept the float open and allowed air to be sucked in through the top of the valve. When the float submerged, the ball was forced back up into its seat and the valve closed to prevent water from entering the snorkel system. (Author's collection)

Figure 5. Snorkel mast in operation. (Author's collection)

All diesel-powered U-boats required air to run their diesel engines. The crew needed fresh air to breathe. If a U-boat attempted to run their diesel engines submerged without a continual source of fresh air, the available air inside the U-boat would be sucked into the engines. The crew would experience excruciating pain due to the vacuum, followed by death through suffocation. Through the use of a snorkel apparatus a U-boat could suck in fresh air while running the diesel engines during a submerged cruise. The boat could be ventilated and the high-pressure air flasks charged. The snorkel as designed allowed for a U-boat to maintain a submerged speed of 6 knots, though some U-boats reached a speed as high as 8 knots. One of the main achievements with the snorkel was that the batteries could be charged while submerged.

Each diesel U-boat type required a slightly different snorkel configuration but the main characteristics were similar. On diesel boats the snorkel apparatus was a streamlined cylindrical mast that measured 9m (30ft) in length. It contained two air conduits. One brought air into the engine compartment of the U-boat through a float valve at the top of the mast designed to stop water from entering the system if submerged. This float valve, also termed the snorkel head, sat about 1m (3ft 3in.) above the water during normal operation. The other line vented the carbon dioxide and carbon monoxide exhaust, oxyhydrogen from the batteries, and other internal gasses from the diesel engines at a point 1m (3ft 3in.) above the top of the snorkel float valve at a 90-degree angle. The design expelled the gas underwater while snorkeling to minimize not only detection, but also interference with periscope vision. The snorkel was raised from a submerged depth of about 25m (82ft) using a hydraulic piston (earlier versions employed a pulley). Once raised, the snorkel was held in place by a staying rod that locked on the mast by a mechanism in the control room. When lowered, the mast was locked in a recessed well on the deck on diesel boats. This was the basic configuration of the snorkel. How it was installed on to each U-boat Type was different.[1]

There were two types of snorkel installations on the Type VIICs. The initial design, known as the Type I, utilized external intake trunking that ran along the portside of a U-boat into the diesel engine. A spring-loaded flange at that opening was seated against another flange on the snorkel mast when it was raised. The very first Type I masts utilized an exposed intake and exhaust line that were clamped together and raised and lowered by a tension based pulley system. The Type I was improved upon when it went into full production by encasing both the intake and exhaust lines into a single streamlined mast that was raised by a hydraulic piston. By the summer of 1944 a new improved version known as the Type II eliminated the problematic flange and ran the induction trunking below deck through the snorkel boot heel. *U-1105* was equipped with the production Type I flange snorkel mast with a hydraulic piston lift. The exhaust piping on the Type I exited the mast by splitting into two separate pipes that ran around both sides of the snorkel mast's flange intake valve before rejoining and expelling the toxic gas out the exhaust portlet below the snorkel head.

At the same time that the Type II mast was introduced, snorkel heads that remained exposed above the water's surface to take in air were covered with an antiradar coating designed to reduce its signature against surface and airborne radar. *U-1105* does not appear to have ever received this coating as its snorkel installation happened very early in the retrofit process, despite its first operational war patrol occurring so late in the war.

The snorkel exhaust trunking on every single Type VIIC and VIIC/41 ran above deck without exception. It departed the diesel engine room, popped up above deck on the starboard side of the conning tower where a shut-off valve was located, before the trunking dropped back below deck near the magnetic compass housing and connected into the snorkel boot heel. The shutoff valve to the exhaust trunking was operated by a turn-wheel in the control room. It is not entirely clear why the exhaust trunking on the Type VIICs ran above deck as no documentation has been located to-date that describes this engineering decision. Wartime technical diagrams also do not reveal a specific reason. It simply may have been routed above deck due to space constraints. However, multiple surveys by the author of the U-1105 wreck site have not revealed an obvious impediment to running the trunking below deck.[2]

A post-war report prepared by a former member of the Kriegsmarine medical staff provides unique detail on the impact of the snorkel on U-Boat crews and the importance of functioning exhaust trunking on the Type VIICs. Dr. Guenther Malorny served as the primary carbon-monoxide tester of the snorkel-equipped U-Boats from the fall of 1943 through May 1944. He authored "Carbon Monoxide on U-Boats".

Dr. Malorny concluded that before the introduction of the snorkel there was no concern of any CO poisoning onboard U-Boats. No one suspected what the impact was until the first testing began in the fall of 1943. It was determined that when operating a snorkel-equipped U-Boat at periscope depth a greater counter-pressure of the exhaust gasses had to be overcome for discharging the diesel exhaust gasses than was required when operating the U-Boat on the surface. The negative pressure inside the U-Boat, which was permanently present during submerged operation, dropped rapidly when an operating snorkel head was submerged. Conversely, the pressure inside the exhaust trunking increased significantly. The deeper the U-Boat submerged while the snorkel system was running the greater the pressure differential. The MAN diesel engines on the Type IXs generally proved capable of handling this differential, but the Germaniawerft (GW) engines on the Type VIIC could not.

The buildup of toxic CO in the engine room occurred during tactical evasion when the U-boat submerged while the diesel engines were still running during snorkeling, as well as during normal switching to the electric motors at the end of a snorkel run when the diesel engines were on idle. In the latter situation, when the GW-engines on the Type VIICs idled for just a few seconds they were not powerful enough to counteract the pressure that built up in the exhaust line. If the exhaust valve was not shut with the retraction of the snorkel mast in time, seawater rushed into the line and starboard engine forcing toxic CO gas back into the engine room.

Normal CO exposure to crews on non-snorkel equipped VIIC and IXC U-boats was measured at 0.013-0.038% of oxygen. On a VIIC snorkel-quipped U-Boat equipped with GW-engines this increased during normal snorkel operations to 0.08-0.12% of oxygen. But when the snorkel head dipped below the sea surface due to inclement weather or a tactical maneuver while the diesel engines continued to run, as well as when they were on idle, this could increase to 0.4-0.7% or higher, which could be fatal.[3] It was determined that a high concentration of CO would induce acute poisoning within 15 minutes and a loss of consciousness or death in 45 minutes.[4]

Special Experience with "Schnorchel"-Equipped U-Boat No. 4 was soon issued during the testing in the fall of 1943 that directed that "if smoke is present in the Diesel engine room for a short period and if this lasts longer than 5 minutes, work has to be done there with rescue apparatus and protective goggles." It went on to say that if prolonged exposure continued than the Diesel Engine Room had to be sealed off and the "...boat must be ventilated thoroughly with fresh air" after surfacing.[5] This order made sense during training, but not during a war patrol as experience would prove. In the above context, the outer exhaust valve formed the only barrier against outside water pressure. The tightness of the outer exhaust valve was a prerequisite for "perfect snorkeling."[6]

Steps were taken based on Dr. Malorny's recommendations after the CO trials to modify the snorkel exhaust trunking by introducing new spring clutches and develop special snorkel-cams for the engine.[7] Unfortunately due

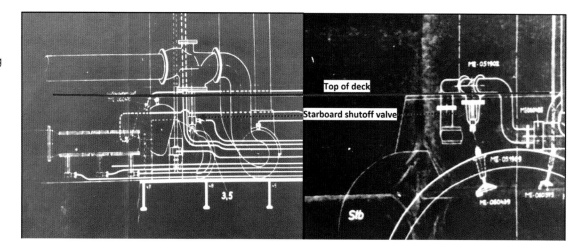

Figure 6. Starboard side exhaust trunking (side view) and shut-off valve (front view, looking aft). These images are derived from wartime German technical diagrams that illustrate the starboard side exhaust trunking present on all Type VIIC and VIIC/41 models retrofitted with a snorkel. This trunking ran above deck on all Type VIIs. This trunking and its purpose went unknown and undocumented since U-1105's initial archeological survey due to a lack of available technical documentation related to snorkel retrofits. It remains a prominent feature still visible on U-1105 today. (Author's collection)

to the prefabrication of snorkel kit production it is unclear how may Type VIICs actually received the new devices that allowed the mechanical blower to be hooked up to the exhaust trunking and keep the outside pressure at bay. By the time the CO testing was completed in December 1943 more than half of all snorkel kits had already been ordered and were on their way to frontline U-Boats to be retrofit. The new snorkel-cams, otherwise known as "camshafts", that drove the diesel motors and were supposed to increased output to overcome the back pressure, were also delayed due to other production priorities and few likely were introduced into the snorkel-retrofit process. However, U-Boat crews soon found work-arounds. For example, a fix was introduced by the crew of U-671 (VIIC), whereby a connection into the exhaust system was arranged from the high-pressure airline to the exhaust manifold. This allowed the U-Boat to economically blow all water out of the exhaust side of the diesels before starting up.[8] Based on a lack of archival available documentation it is not clear if this fix was universally applied across all Type VIIC U-Boats not equipped with the special spring clutch.

Despite this uniqueness of the exhaust trunking design to the Type VIIC (including VIIC/41s), this feature was never identified in the 1993 survey, despite its singular importance to the successful operation of the snorkel. The reason for this lack of identification was simple: not a single published technical source on U-boat construction in any language had identified this feature to date.[9] Only through extensive archival research and photographic analysis has this feature finally been identified and its critical importance in Type VIIC snorkel-equipped U-boat operations revealed.

Four snorkeling operations were generally conducted:

1. Snorkeling while running both engines on charge and propelling the U-boat with silent motors from the battery.
2. Running one engine and using its associated main motor (as a generator) to carry a float current through the control board that was used to power the other shaft. This was known as an "electric drive" by the crew.
3. Straight propulsion in the normal manner for surface operation except that air intake and exhaust was by snorkel.
4. Charging with one engine and direct propulsion with the other engine.

The snorkel device proved more successful than anyone in the German Navy imagined. Initially, it was envisioned that a U-boat would stay submerged only for a few days of operation. It was not known at the time what the physical or psychological effects of prolonged underwater cruises might have on the crew. By the fall of 1944, U-boat crews had pushed the limits of this new technology on their own and were now conducting underwater cruises in excess of 60 days, never surfacing a single time. This was a record that was not broken by the US Navy until the 1970s by the nuclear powered *Seawolf*.

Figure 7. This rare wartime image of *U-1105* taken before its surrender reveals the snorkel mast and head valve. It can be determined from the photo and cutaway that *U-1105* was equipped with a ball float that did not have any antiradar coating, otherwise known as a waffle style "Wesch" absorber. It was equipped with the dipole radar detector. It is likely this photo was taken by a British Sunderland aircraft that flew overhead on May 9, 1945 as *U-1105* maneuvered to Holy Loch to surrender. (Courtesy of U-boot Archiv)

Ladislas Farago, who served as the Chief of Research and Planning in the US Navy's Special Warfare Branch (OP-16-Z) within the US Tenth Fleet during World War II, focused his time extensively on the U-boat threat.[10] Writing after the war, he offered how unprepared the Western Allies were in the face of snorkel-equipped U-boats. The Allies had neither tactics nor technology to counter the new threat, which was the responsibility of the US Navy's Tenth Fleet's "Hunter-Killer" Task Forces in the North Atlantic. Farago wrote:

> In a very real sense, then, the snorkel thus succeeded in doing exactly what Doenitz hoped it would accomplish: it provided effective protection from the U-Boats' most dangerous foe, the planes of the escort carrier groups. The protection was so effective, indeed, that from September, 1944, through March, 1945, the escort carrier groups managed to sink but a single U-Boat, and a non-snorkeler at that, although they accounted for forty-six U-Boats during the prior sixteen months.[11]

U-boat operations evolved. With U-boats staying submerged, communication with BdU was limited or non-existent. This meant that coordinated wolfpack tactics were no longer possible. Instead, U-boats were now being sent to patrol against the shallow coastal waters of Great Britain, Ireland, Iceland, and Canada in order to catch Allied merchant vessels transiting ports of embarkation and debarkation, as well as narrow passages. These operations did not require the high degree of coordination of a wolfpack. U-boats now began to operate as 'lone wolves'. Operating in shallow coastal water meant new tactics had to be developed. Gone were the days when a U-boat would dive deep in the open water of the Atlantic to avoid detection by Allied sonar or incoming depth charges.

BOTTOMING

The main tactic developed by the U-boat force after the introduction of the snorkel was "bottoming". It was recognized early in the war that a U-boat resting on a rocky bottom could likely avoid Allied surface search sonar, as it was very hard for Allied sonar operators to pick out a U-boat in such conditions. However, this tactic was not practiced regularly before 1944 and rarely used up to that time. By the fall of 1944 that had changed.

Bottoming could be employed both offensively and defensively. As snorkel-equipped U-boats were sent to patrol off narrow coastal channels and inlets they could wait on the bottom until they heard a passing ship, then slowly rise to launch a torpedo. This tactic was first broadcast to all U-boat commanders by BdU on July 1, 1944 in an Experience Message 113 titled "Silent Trim".

Resting on the bottom also made it very hard for Allied surface vessels equipped with sonar to locate a U-boat. Toward the end of the war, U-boats were told that after they attacked a vessel, or if they were being hunted by an Allied surface ship, they should immediately "do the opposite" of what the Allies expected and move close into the coast and bottom in shallow water to avoid detection.[12] The Allies grew accustomed to U-boats operating in deeper water and they did not believe a U-boat could operate effectively in the shallows. An entire year of snorkel-equipped U-boat operations proved this belief incorrect. U-boats took full advantage of this misconception and began operating routinely in water depths of 60m (200ft) or less. Often they bottomed in 30–35m (100–125ft), or even shallower, right near land, in order to wait out their pursuers.

While a U-boat still could not stay submerged indefinitely before its air ran out, bottoming allowed the crew to rest as the boat remained completely motionless. They no longer had to actively move about to operate controls or machinery in order to maintain depth or maneuver. The crew essentially made no movement, reducing oxygen intake and extending the length of time a U-boat could remain underwater. Allied surface vessels rarely found a bottomed U-boat, but when it did the chances of a U-boat surviving an attack were high. Depth charges were not as effective against a bottomed U-boat as one higher in the water column. Furthermore, the sea bottom reduced much of the concussive power that was designed to crack open a U-boat's pressure hull. Short of a direct hit from the late-war Allied Hedgehog rocket that detonated on contact, a U-boat often maneuvered away slowly, just feet off the bottom.

When a U-boat had to surface, it raised its snorkel and began to refresh its air. The snorkel head made a very small target for the Allies to locate, especially at night. Radar returns were negligible, especially when the snorkel head was coated with an antiradar coating. Once the air had been refreshed, the U-boat bottomed again at a different location. If the Allied search vessels had departed, the U-boat could turn on its diesels with its snorkel deployed and maneuver away from the area quickly while the Allies continued to search at some other location.

With U-boats now operating closer to the surface due to the need to snorkel, as well as closer to the coast, German naval engineers reinitiated an early war program intended to reduce a U-boat sonar signature that had been previously halted.

ALBERICH

The coating of U-boats with rubber to reduce their sound signatures is one of the least understood German wartime technical developments. In the same way that the snorkel was conceived as a means to counter Allied airpower and radar detection in 1943, acoustic camouflage through the use of a rubber coating had been pursued since the start of the war to counter the sound waves emitted from Allied sonar. Allied sonar, otherwise known as Allied Submarine Detection Investigation Committee devices (ASDIC), was the single greatest threat to U-boats at that time. German technical research in the area of acoustic camouflage was groundbreaking, though its operational employment was hampered through 1942 by a lack of suitable adhesive. By 1943 this had been overcome practically, but its usage was delayed again, only to be adopted with the introduction of the snorkel as it was realized that U-boats would have to remain submerged closer to the surface and for longer periods of time.

The name of this coating was actually "Alberich Verfahren", which was an allusion to the Wagnerian *The Ring* cycle dwarf who could make himself invisible by means of a cap. In the end, the code name was shortened to just "Alberich".

German engineers understood that the value of acoustic camouflage was inherent in a submarine's ability to "remain uninterruptedly underwater" given the advanced state of Allied radar. It was the introduction of the snorkel that provided this ability to U-boats in 1944, thus linking these two technical innovations.

Alberich was a two-ply rubber coating with intermittent holes that had to be uniquely patterned for different areas of the U-boat. The overall performance of Alberich was dependent upon both pressure and temperature. This necessitated the preparation of designs to meet varying operational requirements. As the pressure (depth) increased, the numbers of both large and small holes had to increase in order to maintain optimal conditions. When the depth decreased, the opposite was required. What this meant was that if a U-boat was configured with an Alberich coating to function at a depth of 40m (130ft) or less, but found itself in an evasive tactical situation at a depth of 150m (492ft), the reflective reduction of its coating would be greatly diminished, making it more visible to ASDIC detection. Temperature, however, had an opposite effect, so that the reflection loss factor decreased with increased temperature, requiring a smaller number of holes to meet the operational conditions.[13]

Early in the war, the initial operational requirement from BdU was to tune Alberich to function at a depth of 150m (492ft). The original thinking was that if under an attack by surface vessels, a U-boat would attempt to depart the area by diving deep. In theory, the Alberich coating would reduce the U-boat's signature enough to prevent the surface vessel from zeroing on the U-boat as it made its escape. However, this requirement changed in 1944 and Alberich was tuned for a depth of 40m (130ft) or less. This change was made for two reasons. First, the initial tuning of Alberich for 150m (492ft) was defensive in nature. It was then subsequently believed that tuning Alberich to be effective at 40m (130ft) or less would mask its

Figure 8. A rather clear photo of *U-1105*'s Alberich coating along the portside bow. The Alberich pattern varied along different parts of the hull due to do the calibration required for different hull thickness, water or air backing. In this photo several different patterns can be discerned between the Alberich affixed lower and higher on the bow. (Courtesy of Maryland Historical Trust)

Figures 9–11. All three pictures illustrate Alberich recovered from *U-1105*. Figure 7 shows the variant nature of the application as the left piece was recovered from the conning tower while the right piece from the saddle tank. The two separate photos of Alberich that appear to come from the conning tower currently reside at the Piney Point Lighthouse Museum (left) and the U-Boot Museum, Germany (right). (Courtesy of Maryland Historical Trust and author's collection)

approach to a target, enabling surprise in an attack. Second, the introduction of the snorkel meant that U-boats were now going to operate for long periods at shallow depths, solidifying Alberich's shallower application.

The laboratory development of Alberich resulted in a two-ply rubber sheet 4mm thick cemented on to the hull and superstructure of the U-boat. The inner sheet was perforated with 2mm- and 5mm-diameter holes that determined the resonant nature of the coating. During Alberich's development phase German engineers confirmed the correct tuning through a process whereby the two sheets were energized by passing an electric current through respective hull plating. How the Alberich absorbed the current was then measured against the principal parameters of hull plate thickness and whether the backing was air or water (as in the case of the ballast tanks). The result was that on a standard 6mm outer hull, the number of perforations required per each 20sq cm/3.1sq in. piece of Alberich was 53 x 2mm and 11.6 x 5mm holes. The thicker the hull plating the more 2mm holes and fewer 5mm holes were required. The thinner the hull plating, the inverse was true.

Initial trials confirmed that Alberich could reduce the effectiveness of Allied sonar in the 10–18kHz range by approximately 15 percent, but not consistently through varying depths. This led to the concept of tuning the acoustic camouflage by adjusting the number of 2mm and 5mm perforated holes applied to any given 20sq cm/3.1sq in. sheet, as noted above.

The initial sea tests were conducted in the Skagerrak strait against a coated U-boat by a surface vessel traveling at 7 knots. These tests resulted in an inability to verifiably locate the U-boat, while an uncoated U-boat was detected at a range of 1.6km (1 mile). Knowing that the limitation was the noise of the coated U-boat, whether it was running on electric or diesel engines, the tests proved that a reduction in reflectivity of 20 percent would reduce the range of ASDIC detection anywhere between 20–60 percent. This was a marked advantage against Allied ASDIC.

Alberich proved effective as acoustic camouflage during the war given the fact that no U-boat covered with the rubber was identified by ASDIC or sunk by a surface vessel. However, only 13 U-boats were covered with the final operational version of Alberich. Of those, only four saw an operational patrol. Two were sunk: one by a fixed underwater mine in the English Channel, and the other by a British submarine that caught it on the surface just outside its Norwegian base as it was returning from a patrol.

Alberich can still be seen along *U-1105*'s conning tower and exposed portion of its saddle tanks. A quick wipe of the sediment with one's gloved hand easily exposes the black rubber coating.

GHG BALKON

Both the snorkel and Alberich were conceived of initially as defensive developments, though they both allowed a U-boat to maintain offensive viability against the Allies through the end of the war. Now that a U-boat was operating almost exclusively underwater it required improved listening equipment to decipher the various sounds in shallow coastal waters that could interfere with the identification of a surface vessel, especially while operating the snorkel.

As early as mid-1942 German engineers designed a modification to the existing standard placement of the passive sonar array known as *Gruppenhorchgeräte* (GHG). The standard GHG was placed along the starboard and port sides of a U-boat's bow, just behind the torpedo doors. They proposed to reduce its complexity and increase its effectiveness at shallower depth through the installation of 12 hydrophone receivers on each side of the keel. Installation and testing was conducted in the summer of 1942. The new placement improved detection at shallower depths when running on electric motors as previously the sounds of waves had interfered with the acoustic sensors. With the new device the U-boats gained improved listening close to the surface, but they had to maintain a depth of about 20m (65ft); any shallower and the acoustic sensors simply did not work well. Later, when the snorkel was introduced, a U-boat had to occasionally stop its engines completely and turn off the snorkel in order to quickly check the sonar and determine if any enemy vessels were approaching.

Further testing resulted in a new proposal to develop a *Balkon* or semi-circular "balcony" that would extend beyond the front of the keel and be equipped with 2 x 24 receivers. By September 1942 a new version had been tested and returned better than expected results. Overall the *Balkon* increased the interception range by 70 percent when compared to the near frontal placement of the original system. This increased the forward sensing capabilities of the GHG system, though it left a blind spot aft between 150 and 210 degrees.[14]

The first two U-boats that received the GHG Balkon in the late fall of 1943 and early spring of 1944 had not yet been retrofitted with the snorkel. One U-boat was sunk and the other captured at sea before a retrofit could occur. After that, all U-boats that received a GHG Balkon were equipped with a snorkel.

After the introduction of the snorkel it was determined that sound bearings while snorkeling and running the diesel engines caused considerable issues with the standard placement of the passive GHG sonar array. Sonic information could not be received with accuracy, if at all, when a U-boat was submerged running the diesel engines while snorkeling. The new GHG Balkon, however, would overcome many of these listening challenges. The larger Type IX U-boats thus began to receive this device in the spring and summer of 1944, followed by the Type VIICs in the fall of that year.

Approximately 22 U-boats received this device and *U-1105* was among that last to have it installed, possibly in early 1945. There may have been other U-boats that received the GHG Balkon, but the historical record is not complete in this regard.

U-1105 could count itself among a number of very few U-boats equipped with these three technologies and trained with the new tactics to employ them. It was the only U-boat so equipped to conduct a wartime patrol. Unfortunately, its GHG Balkon apparatus is buried in the mud and cannot be seen.

Figure 12. GHG Balkon on *U-1105*. This advanced passive sonar array gave *U-1105* a significant advantage in detecting Allied vessels. *U-1105* was the only U-boat to conduct a combat patrol equipped with a snorkel, Alberich, and the GHG Balkon. (Courtesy of the Royal Navy Submarine Museum)

CHAPTER 2

CONSTRUCTION AND TRAINING

U-1105 was built by the Nordseewerke Shipyard in Emden. It was the 25th U-boat constructed by the yard. The keel was laid down on July 6, 1943 and it was launched on April 20, 1944. On June 3, 1944 *U-1105* was officially commissioned as a VIIC into the German *Kriegsmarine*.

U-1105 has unfortunately been misidentified in a number of published and online sources as being a Type VIIC/41. The only difference between a Type VIIC and a Type VIIC/41 is that the latter was designed to have a slightly thicker pressure hull in order to dive deeper in the open Atlantic Ocean and avoid Allied depth charges – a tactical benefit never realized as U-boat operations shifted to shallow coastal water before the end of the war. This error can be traced to Günter Hessler's history *The U-Boat War in the Atlantic 1939–1945*.[15] Written while being held captive by the British, Hessler was given access to available German records in order to compile an operational history of the U-boat force from a German perspective. In Appendix II of Hessler's book he incorrectly lists U-boat hull series 1103–1110 as Type VIIC/41. This annotation was likely a simple compiling error. It should also be noted that the British only gave Hessler a limited number of documents on which to base his history, not everything that was available. A more reliable document is the US Navy's postwar Technical Report Number 312-45 "German Submarine Design, 1935-1945". This report relied upon all available German wartime documents to include dockyard construction records of U-boats. It lists U-boat hull series 1101–1106 as a Type VIIC. The confusion on the part of Hessler might have been caused by the fact that with the exception of 1101–1106, hull numbers 995–1110 were all designated VIIC/41.

A further confirmation of *U-1105*'s Type comes from its construction *Shiffsbuches* located at the Bundesarchiv-Militärarchiv in Koblenz, Germany.[16] This was a standard form document issued at some point between when the keel was laid and the hull launched. It contained technical characteristics and a list of the standard fittings and those that were intended to be installed on each U-boat. In the case of *U-1105*, the *Shiffsbuches* notes that the thickness of the pressure hull's sheet metal was 18.5mm. This is the exact thickness of a Type VIIC. The only difference between a Type VIIC and a Type VIIC/41 was that the latter had a slightly thicker pressure hull of 21mm. There is therefore little doubt that the plan was to construct *U-1105* as a Type VIIC U-boat. Perhaps final confirmation of archival evidence may occur with the measurement of *U-1105*'s pressure hull during a future maritime archeological survey of the wreck. Until such a time, *U-1105* should be classified as a Type VIIC based upon available archival documentation.

When *U-1105* was ordered and the keel laid in 1943 the U-boat war was in transition. The *Shiffsbuches* reveals that *U-1105* was initially to be equipped with the standard configuration of an 8.8cm SK deck gun forward and a 2cm Flak C/30 aft on what was termed a Turmumbau II – better known as a Wintergarten – prevalent in 1942–43. It was to be equipped with a standard GHG sonar array, though this was later modified to the advanced GHG Balkon version during a final overhaul. There was no mention of either a snorkel or Alberich.

The Battle of the Atlantic began to evolve starting in the fall of 1943 as a result of "Black May", and so did the U-boats. Dönitz introduced a number of changes and technical solutions designed to improve his boats' survivability. Thus, before the completion of construction, *U-1105*'s 8.8cm deck gun was eliminated, an expanded Turmumbau IV, multilevel antiaircraft gun platform aft of the conning tower was added, two twin 2cm Flak 38 cannons were placed on the upper deck, and a 3.7cm automatic M42 was fitted on the lower deck. These changes reflected the need to better protect the U-boats against Allied aircraft along with the snorkel that began to enter service late in 1943. *U-1105* was among the first U-boats designated to receive both a snorkel and a coating of Alberich before being commissioned.

U-1105's Alberich coating was applied in Kiel between October 1943 and January 1944. The snorkel installation was a Type I flange with a ball float. No antiradar Wesch matt was applied to the snorkel head and upper mast.[17] An aerial photo of *U-1105* taken during its surrender at the end of the war reveals an uncoated ball float attached to a snorkel mast recessed in the deck well. Admiralty documents relating to the following testing also appear to confirm that no antiradar covering was applied to *U-1105*'s snorkel head and upper mast.

During its final overhaul, a new GHG Balkon passive sonar was added to the underside of the U-boat below the keel that increased the interception range by 70 percent when compared to the near frontal placement of the original system. It was also equipped with the FuMO 61 Hohentwiel radar and the Wanze 2 radar detector, though snorkel-equipped U-boats rarely – if ever – used these devices as they spent almost no time on the surface.

U-1105 did not have all the innovations of U-boats launched in the fall of 1944 or early 1945. For example, it did not have automatic depth-keeping gear that some snorkel U-boats received.[18] Yet, according to the U-Blatt available in the U-Boat Archive at Cuxhaven, which was likely completed with the input from former crew members after the war, *U-1105* was possibly equipped with the 'Kurier' flash transmission system, though this remains unconfirmed by a primary document. Flash transmission was first tested in the spring of 1943. It compressed a communication, which could contain as many as seven letters in the Morse alphabet, into an elapsed time of 425 milliseconds. This not only prevented the U-boat from being located through radio direction finding, but also prevented burst transmissions from being intercepted and read by Allied cryptologists. The system contained two parts. A Phillip CR-101-A Receiver and a 'Geber' KZ G44/2 (Pulse Giver) transmitter. Snorkel-equipped U-boats began to be equipped with the flash transmitting gear in the late fall/early winter 1944. The only drawback was that it had limited range. The presence of this new radio system might offer one reason for the lack of *Ultra* intercepts for this U-boat during its brief patrol. Arguably, snorkeling would also eliminate radio intercepts as U-boats could not communicate when submerged or while employing a snorkel. Most U-boats equipped with a Kurier system threw their transmitter overboard before they surrendered.

Nevertheless, *U-1105* was as advanced as a U-boat could be when it was launched in 1944. It boasted the latest electronics, a snorkel, Alberich coating, and GHG Balkon. It joined a small group of similar U-boats that represented a clear shift in operations. While it did not have the strengthened hull of a VIIC/41 that would give it the ability to dive deeper in the mid-Atlantic, *U-1105* did not need that. It was going to operate in shallow coastal waters.

What follows is an annotated history of *U-1105*'s wartime experiences as written by the U-boat's captain Klt Hans-Joachim Schwarz. His account has been combined with assorted primary documents that include *Ultra* intercepts and the war diary of the Royal Navy's 21st Escort Group.

On June 3, 1944 Klt Schwarz stated to his men:

Comrades!

Today, we have finally come to the day we all awaited for a long time, the day that already seemed within reach but then was pushed into what appeared to be the distant future, the day that we were going to be able to put our boat into operation.

In months of tedious work, the shipyard built this boat that will be our homeland from now on. I would hereby like to thank the shipyard for the work they have done and also assure it that we will always value their work and that we want to make the boat into a powerful fighting instrument.

> Effective today, we will be included in the great society of the German U-boats force. It is a great honor for us to be members of a branch that did such shining work in the last World War and has achieved such powerful results in this war. It is, however, also our duty to be the shining examples of the German submarine branch to the men and their courageous crews that men like Weddingen, Saltzwedel, Lohs, Prien, Schepke and Endraß were and to show ourselves worthy of them. It will be the job of us, the young submarine crews, to continue the fight that these men have started so successfully and fight our way through to the bitter end.
>
> Even though at the moment it has become a bit calm as concerns the U-boat arm, that it no reason for us to hang our heads. We as soldiers can never lose our faith in our weapons, because anyone who does not believe in his weapons will never be successful in using them. But in the words of our Commander in Chief, the submarine branch will someday be the branch that had the significant role in deciding this war. And we can rely on that.
>
> So we know want to take over our boat in a belief in our weapons, trusting our commander, and in unshakable faithfulness to our Führer!
>
> "Crew halt, face the flag!" – By order of the Führer and the Commander in Chief of the Wehrmacht, I put *U-1105* into service. Raise the flag and the pennant."
>
> "To our Führer and the Commander in Chief of the Wehrmacht, Sieg Heil, Sieg Heil, Sieg Heil!"

According to Schwarz, "three days after commissioning, *U-1105* left for Kiel on the day that Emden was bombed by RAF aircraft."

Before we got to the lock, we almost rammed into a train of barges. But at the last minute, God Almighty put his thumb in between us. Maybe it was the First Watch Officer having the presence of mind to

Figure 13. *U-1105* on the day it was commissioned, June 3, 1944. Klt Hans-Joachim Schwarz is standing on the apex of the lower Wintergarten. (Courtesy of Maryland Historical Trust)

jettison the bouquets that we had been given at our farewell that saved us. And that was how our first ocean trip started!

The trip through the canal went off in marvelous weather. When we passed the Reich Colonial School in Rendsburg, our boat as well sent over its traditional greetings,

"June 8 – June 27 UAK [U-boat Acceptance Commission] in Kiel.

Diving, attempts to trim, snorkeling, firing torpedo models, compensating, radio correction, etc. One trial followed another, frequently interrupted by an air-raid alert. But in spite of obstacles, buoys and thick artificial fog, *U-1105* always found the protective submarine bunker.

An *Ultra* intercept confirmed that *U-1105* conducted its first successful snorkel acceptance trials from 11:24am on June 13, 1944 to 1:31pm on June 14, remaining under water for more than 24 hours.[19]

There was no lack of amusing episodes in Kiel. While the First Watch Officer was trying to find the depth of the Kiel harbor elegantly, Bootsmaat Erdman was trying to do that with his wristwatch. The voluntary diver unfortunately was unsuccessful; he even left a lead shoe "below". After attempting to listen in on June 27 outside of Sonderburg, we started off toward Swinemünde.

U-1105's movement was confirmed by an *Ultra* intercept at 2:19am on June 28 when a message was transmitted from *U-1105* to BdU with the statement "Eastbound passage."[20]

From June 28th to June 30th, there was a proper air defense training session at Air Defense School VII, in which seamen and firemen participated. On July 1st, we left the air defense school and

Figure 14. Schwarz is pictured on the right in a white cap. To the left is Obersteuermann Fröhlich, the Navigation Officer of *U-1105*. This picture was likely taken during training in the Baltic. (This picture was given to Henry Keatts by Schwarz in the 1990s. Courtesy of Maryland Historical Trust)

while the weather was good, we started our trip to Danzig. One day later, the boat tied up to the UAK pier in Kaiserhafen. The residential ship *Iberia* was in front of us.

Danzig! A concept for everyone traveling on a submarine that is associated with many lovely stories. Words like Tobis, the Heubude beach and Zoppot spoke to each of us and called forth the most beautiful memories associated with them when we heard them.

Off the coast of Hela, the first deep dives and long-distance trips were done. From July 4th to July 27th, we unfortunately had to look for a "Holm Fix-it Shop" because of a small battery problem. But even so, no one regretted the four-week stay in Danzig.

On July 28th, we went to the 19th Submarine Flotilla in Pillau, to do dry tactical escort training, aircraft ID service and general boat maneuver and travel exercises there. Individual training in the Gulf of Danzig was included in it.

From Pillau, we wound up in Hela on August 4th, 1944. "Hela" – the name resonates in the submarine branch! Solitude and a tight schedule are the characteristic features of this training facility. Under the leadership of a much-traveled "Commissar" the boat was wound up in diving quadrants and areas of operation in five exercises. Rivers of sweat engulfed our foreheads. The most cunningly-devised breakdowns were added. The call from above and frequently from the control room as well, "Lord, let it be evening!" was taken up. Everything was done according to the motto, "hard but not unfair." For this motto, everyone became accustomed to the most impossible burdens. The much-discussed snorkeling mast was also tested. Pressure differences up to 500 mb pulled out a lot of men's lugs further out. A day of rest in Gotenhafen was used for an on-board party.

A communiqué that was intercepted by *Ultra* went to BdU announcing *U-1105*s arrival in Danzig where it joined the 8th U-Flotilla.[21]

From August 17 to September 5, 1944, the boat's group front training was interrupted due to renovation work and was back in the Holm shipyard in Danzig.

On September 11 at 11:00pm *Ultra* noted that while in Danzig, Schwarz requested permission to dive in order to test the Alberich coating.

This *Ultra* intercept is handwritten and not typed. Apparently the request was refused, as noted on the intercept, due to concern about air-dropped mines.[22]

Then the training was continued in Hela and on September 13th, 1944, ended with the award of the red triangle.

The group front training was joined by the pre-tactical training at the 20th U-Flotilla from September 15th–23rd. The entire mid-Baltic Sea was covered. Here flight training, attack training and movement exercises alternated with tasks on maintaining contact and escort platoon. In the process, high demands were placed on running the ship, steering, the bridge personnel, machinery and running the radio. When traveling in a group, our neighbor boat exercised such attractive force one day that the "Black Panther" could not avoid rubbing against it. With slight chafing on both sides, however, we managed to get apart without much damage.

Misfortunes are rarely alone! For example, when we came into Danzig at night on September 27th, no one thought that the camouflaged pier might be slightly lower than the extended periscope.

Because it was the commander's birthday, the matter could be "adjusted" back the next day with a schnapps and a few cigars.

From October 2nd–21st, *U-1105* was in the dock at the Deutsche Werke in Gotenhafen.

On October 23rd, we went on to Pillau, where TEK [Torpedo Test Command] tests were done and then we did firing training from October 27th to November 6th with the 26th Submarine Flotilla. There was a lot of work there and little sleep. Firing during the day and the night above and below the water alternated with record times in accepting torpedoes. In the meantime, a small ramming with a K torpedo chase boat could not be avoided, but thank God, all it caused was damage to the sheet metal. At any rate, we booked 85 percent hits as a final result, of which 22 were "center of mass hits."

Until November 11th, there was individual training at sea in which seaman numbers I and II distinguished themselves as particularly good catchers of salmon. With the flotilla flag of the Head of the 26th Submarine Flotilla on our mast, we went back to Danzig on November 14th.

The tactical exercises that started on November 20th were the crowning achievement of our entire homeland training. The Baltic Sea, which was for a time whipped up as blue and green, was covered multiple times from the Gulf of Danzig to Bornholm. One escort group after another was the surface or the underwater target of our attacks. Once again, high demands were placed on the boat and its crew. There was no rest in the radio room. The 2000th radio transmission went over the airwaves at that time. The senior pilot was only seen bent over, linked up, tossing and pondering over his charts, basically on the "it's navigation if you get there anyway!" theory. Thanks to their good care and maintenance, the diesel engines ran without a problem like sewing machines that were almost "Gustav Fritz". Here as well, there was an anniversary to be celebrated: the ten millionth engine turnover. In the tower, one could continually hear the Underwater Torpedo Head yelling "cover, cover," otherwise one could see him in the dining room where he was putting together potato salad – his "specialty". After our boat's attack spirit was depicted as exemplary in a radio transmission from the "leadership" and we were also able to put a "*U-1105* did well" in the radio transmission blotter, into Gotenhafen with our pennants set out.

That ended our homeland training and we were released from training by Kapitän zur See Schütze, the submarine commander, after a formation.

However, before we could start our trip to the Wilhelmshaven shipyard for the remaining work, we went to Hela one more time on November 29th, where we had to take Head Engineer trainees. On December 4th, we finally sailed by Hela for the last time.

This was the first *Ultra* intercept in two months that was identified for *U-1105*. This intercept captured the request for permission to move independently from Danzig to Swinemünde on December 5. Permission was granted.[23]

The next morning, the boat was released from the 8th Submarine Flotilla and was cleared for ocean travel for the trip to Swinemünde. Here we paid one more visit to the air defense school. From morning to evening, "the cap was back on the carriage." After we completed our functional and training firing, we finally set off to the "pearl of the Baltic" [Kiel] on December 13th, and it opened the gates to the lock at 1:00 a.m. on December 15th, 1944.

U-1105 continued toward Brunsbüttel at the mouth of the Elbe River. It was noted that the passage through the Kiel Canal was delayed by fog and that predicted arrival was on December 15.[24] This was the last intercept *Ultra* captured until March 1945.

That meant that *U-1105* had traveled 7,675.5 nautical miles, of which 7,038.5 were on the surface and 636.7 underwater, corresponding to a trip from northern Spitzbergen to Cape Horn.

While the boat was turned over to the naval shipyard for its final work, the crew was able to go on its earned leave during its cruise. In addition, the time for final work lasted much too long. Everyone longed for the moment when *U-1105* would be back diving into its wet element.

U-1105 departed in convoy for passage to Kiel on March 19 and arrived at Buoy A13 on March 23, where it reported ready for multiunit hydrophone trials.[25] On April 3 it reported that onward passage with *U-2328* (XXIII) was delayed by a gale. They reached "Anchorage E" and were awaiting onward escort as reported on April 4.[26] On April 5, *U-1105* and *U-2328* parted ways after being released from their escort.[27]

Schwarz's account continues:

> After undocking in Wilhelmshaven, we went through the Kiel-Wilhelmshaven canal to Kiel to get our final equipment for the enemy trip:
> - taking on food in Tirpitzhafen
> - fuel in Holtenau
> - taking on ammunition at the arsenal in Dietrichsdorf
>
> Interrupted by air attack alters and seeking shelter in the submarine bunker.
>
> In early April 1945, we sailed out of Kiel with *U-2328*, Oberleutnant zur See Lawrence, a type of XXIII boat, through the Great Belt and the Kattegat to the Oslo Fjord.
>
> *U-1105* arrived at Horten on April 6th at 8:24am as reported to

BdU and intercepted by *Ultra*.[28] As standard practice, final deep diving and snorkel training then took place.

We did one last snorkel training session in Horten with a function test of the technical equipment.

U-1105 was now ready for its first and only combat patrol.

CHAPTER 3

FIRST AND LAST WARTIME PATROL

At 12:15pm on April 10 *Ultra* intercepted a report that noted *U-1128* had departed on the 9th for a war patrol and that *U-1105* would depart on independent passage from Horten to Kristiansand South. After some delays, *U-1105* officially departed on its first war patrol at 9:37pm on April 12.[29]

While *U-2328* was going to Narvik, we went to Kristiansand on the southern coast of Norway. Because of heavy English air cover, we could only move underwater. In Kristiansand, we met *U-1228* (IXC/40), [under the command of] Oberleutnant zur See Marienfelde, a type of IX boat with which we then sailed out the next evening after it was dark. A lead boat escorted us to the "Krieger" point near Cape Lindesnes. There we parted ways and *U-1105* set course for Fair Island between the Orkney and the Shetland Islands.

On the 17th, *U-901* under the command of Kapitänleutnant Hans Schrenk, *U-1010* under the command of Kapitänleutnant Günter

Strauch, and *U-1105* all received a "Guard Diana" order, meaning that they had to switch to that radio circuit starting on 8:08am on the 18th.[30] Both *U-901* and *U-1010* departed Stavanger enroute to England's coastal waters.

Our operational orders provided that because of the heavy monitoring of the sea and air, we would go around the Shetlands on the north side and reach the Atlantic by going between the Shetlands and the Faroe Islands. We were assigned a large square off the west coast of Ireland that the escorts from Halifax and Newfoundland, USA [sic] had to pass on their way to the North Channel.

U-1105 was operating independently off the west coast of Ireland near Broadhaven Bay, looking for convoys that were heading north toward the North Chanel. Its assigned area of operations was broadcast on the 21st by BdU; "As attack area occupy SQ Blue of 8123 ((AM 5723: 54.33 N 10.35 W)) with depth of 40 miles."[31] The last reference was likely a "no further away from coast reference of 40 miles," though alternatively it might have been miss-translated from a "no deeper than 40 meters" reference. It was likely the former as BdU rarely issued that level of tactical guidance at this stage of the war.

Because the sea between the Faroe Islands and the Shetlands was also very strongly monitored and many boats had been lost there recently, the commander decided to take the Fair passage in spite of the risk associated with it. The course was plotted for the north passage, so Fair Island had to be passed on the port side. Although the determination of the ship's location could only be done using dead reckoning and possibly depth sounding, and thus was not particularly exact, we saw the island pretty accurately on our starboard side. We could easily recognize the lighthouse and the vehicle traffic on the island through the periscope.

We avoided multiple guard vessels looking with ASDIC by traveling at greater depths. After we had left the Orkneys and the Shetlands well behind us, we moved our course to the southwest at about 60 degrees north, 4 degrees west. We went under a large English minefield area at 100 meters depth. When we had reached the Hebrides at 58 degrees north and 9 degrees west, we moved at a 180-degree course that was supposed to take us directly to our area of operations.

> Now for the first time, we surfaced for a quarter of an hour to send off the first radio transmission with our report location and to push the accumulated kitchen waste overboard. While we were traveling onward, the St. Kilda group of islands, which is in front of the Hebrides, came into view.

This transmission was intercepted by *Ultra*. Schwarz transmitted at 3:39pm on April 23 and stated: "My position is AM 0213 ((57.15N, 09.48W)). Weak defense and patrol."[32] This was the only time that *U-1105* surfaced to send a communication back to BdU during its combat patrol. This put *U-1105* 50 nautical miles west of North Uist along the continental shelf heading south. As per BdU guidance, U-boats traveling submerged used the continental shelf and other known underwater features as a navigational aid to reach their designated patrol area.

Snorkel-equipped U-boats occasionally experienced technical issues with the system. One of the more frequent of these was a stuck ball float that often required the U-boat to briefly surface so the crew could conduct its required maintenance. Such a situation was a dangerous time for a U-boat and they commonly moved very close to the coast, knowing that the Western Allies still did not expect a U-boat to surface in shallow water. Schwarz details such a situation below:

> The trip went without incident until we reached the area of operations. The closer we got to the ocean area near the North Channel, the more we registered propeller sounds from guard vehicles and the distant booms of depth charges on our listening devices. In Broadhaven Bay, right off the rocky coast of Ireland, we surfaced again to do a repair of the snorkel. We were so close to the coast that we could see the lights, advertisements and cars with the naked eye.
> There were no earlier references to any issues with the snorkel and no detail as to what this particular issue was.
> The next evening, the radio room reported propeller sounds from a warship some distance away. We then went to periscope depth and to our surprise, we saw a large two-stack destroyer in the moonlight in an ideal position for firing. Unfortunately we were unable to fire because while we were setting up the torpedo tube, it came toward us and we had to go to the depths.

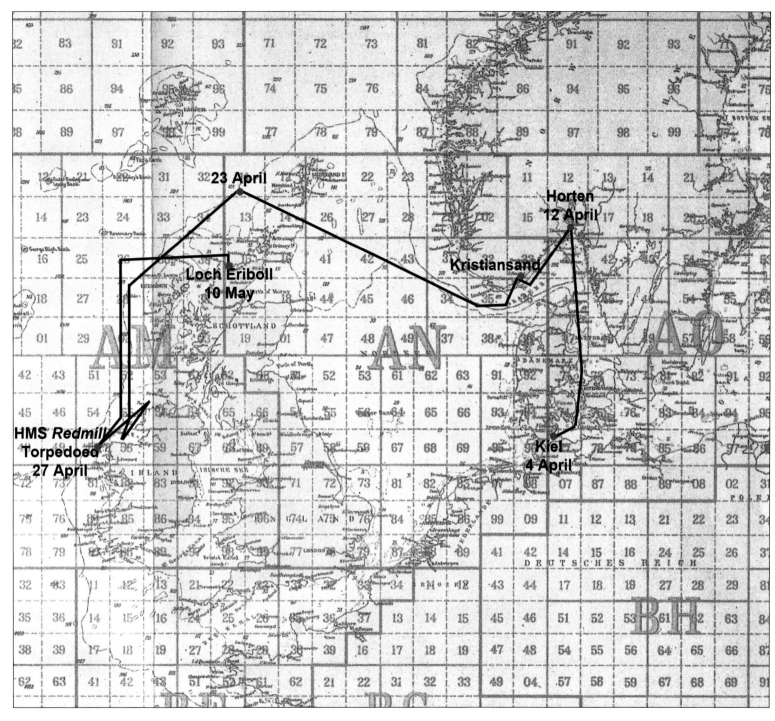

Figure 15 This map depicts *U-1105*'s combat patrol as derived from Schwarz's own rendering. After departing Kiel it arrived at Horten where it conducted its final deep-diving and snorkel tests, as did almost every departing U-boat at that time. Its area of patrol was off Black Rock, Ireland and the approaches to the North Channel. As in the case of all snorkel-equipped U-boats, *U-1105* cruised almost exclusively underwater after departing Norway. In order to maneuver effectively to its patrol area it employed its depth sounder to identify the continental shelf along the 200m (219yd) line as a navigation aid. (Author's rendering)

On the morning of April 27th, 1945 there were more propeller noises reported, and we came into view of a search group consisting of 3 frigates, which we attacked. When we had fired off a Zaunkönig [acoustic torpedo], a frigate suddenly broke out of the group and headed toward us; it had clearly discovered us. At a distance of about 900 meters, the commander, with his prow to the right at an acute angle, fired a LUT [bearing-independent torpedo] in addition as what he called a destroyerdefense shot and took the boat immediately to 100 meters depth.

It is very likely the new GHG Balkon gave *U-1105* the initial edge in this engagement, allowing Schwarz to detect the British escorts before his U-boat's torpedo was detected. No doubt his Alberich also worked as designed, masking *U-1105*'s shallow-water approach toward the surface vessels. As related below, HMS *Conn* of the 21st Group's 1st Division did in fact detect the acoustic torpedo and turned abruptly.

After 50 seconds, there was a detonation – our torpedo had hit something. Shortly thereafter, there was a second detonation; the Zaunkönig had hit its target as well. Whether it was the same frigate or a different one in the group, we couldn't determine.

It is likely that only one torpedo struck the HMS *Redmill* as detailed below.

The sounds of sinking could be heard without a problem, not only in the listening device but also on board with the naked ear. There were then a few more explosions – either frigate boilers or the depth charges on the decks of the frigates. We did not find out which ship it was until we read the book "Chronik des Seekrieges 1939-45" [*The Chronicle of the War at Sea*, 1939–1945] by J. Rohwer et al., Hümmelchen, Stalling-Verlag, which reports the following on page 539:

"*U-218* (Kapitänleutnant Stock) lay a minefield in the area of the Hebrides and the North Channel. *U-636* was sunk by DE Bazely, DE Drury and DE Bentinck of the 4th Escort Group. *U-293* (Kapitänleutnant Klingspor) or *U-956* (Kapitänleutnant Mohs) sunk 1 ship with 878 GRT; *U-1105* (Oberleutnant zur See Schwarz) torpedoed DE Redmill and avoided being located by the sonar of the ship chasing it thanks to its Alberich coating."

The HMS *Redmill* was an escort destroyer with 1,300 tons displacement and 25 knots speed. It belonged to the 2nd Division of the 21st Escort Group.

The remaining boats of the search group followed us with depth charges, but were not successful. In the meantime, *U-1105* was not stopped at the depth of 100 meters to which it was headed; it continued to plunge down all the way through. To stop the boat, the bilge pumps were used at every series of depth charges. The depth gauge read 140, 150, 160 meters – trend downward. At 172 meters, there was suddenly contact with the bottom. The boat moved some more along the soft sand and then we turned off the engines, took on two tons of negative buoyancy and remained right there. A stone had been lifted from our hearts!

In the meantime, our pursuers had received reinforcements at sea and were doing a systematic search using their ASDIC locating system. Because of our Alberich equipment and the fact that we were on the bottom, however, *U-1105* was not picked up. A total of 299 depth charges were counted. A small water leak was quickly remedied. The series of depth charges moved further and further away. After a total of 31 hours – about 4:00 p.m. the following day – we got off the bottom and moved to periscope depth. Because this search group was very far away, we left this area, which was now in danger, and went toward Tory Island. There was little ship traffic there, so after a few days we went back to our assigned area of operations near Black Rock.

The news we got by radio from home reported the end of the war. On May 4, 1945, all of the boats at sea got orders to stop combat. The last order from the Commander of Submarines [on the 8th] read,

"My U-Boat Men!

Six years of submarine warfare are behind us. You have fought like lions. A crushing superiority in materiel [sic] has forced us into an extremely tight spot.

Based on what we have left, continuation of our battle is no longer possible.

Submariners, lay down your arms unbroken and spotless after an unequalled heroic fight. We remember and venerate our fallen comrades, who have confirmed their faithfulness to the Führer and the Fatherland with their deaths.

Comrades, continue to maintain your submarine spirit with which you have courageously and unflinchingly fought for years for the best interests of our Fatherland in the future as well.

Long live Germany

The Grossadmiral"

U-1105 got instructions to surface and move on a prescribed course to the base at Loch Eriboll on the northern coast of Scotland and surrender there. On the way, all the torpedoes still in tubes were thrown overboard, as were all the classified material, the locks to the air defense weapons and ammunition. There was one more tense situation for us when an English aircraft flew at a very low altitude over us, even though there was a prohibition against that. We would have had no chance if we had been attacked by bombs.

Everything noted by Schwarz in his history of *U-1105* is corroborated by primary documents except one incident. Schwarz claimed, and it has been often repeated since it was first published by Henry Keatts in his 1994 book *Dive into U-Boats*, that *U-1105* was pursued by elements of 2nd Division of the 21st Escort Group that dropped nearly 300 depth charges in their attempt to destroy the attacking U-boat. This claim does not hold up to the historic record. The following information is derived from the war diary of the 21st Escort Group that covers the period of April 21 through May 23.

As an introduction to this war diary there is a letter penned by the senior officer of the Group that reads:

Sir,

I have the honor to submit Report of proceedings covering the period April 21st to May 23rd, 1945.

It is with deep regret that I have to report that H.M.S. "Redmill" was torpedoed on April 27th 1945 with the loss of 28 of her gallant ship's company. The conduct of her Commanding Officer and entire ship's company after the ship had been torpedoed was exemplary and in accordance with the highest tradition of the Service. During her career with the 21st Escort Group this ship was outstanding in all respects and I have been proud to have her under my command.

> To be present to take part in the surrender of 33 U-boats at Loch Eriboll was a source of great satisfaction for the 21st Escort Group.
>
> As this may be the last Report of Proceedings which I submit as Senior Officer of the 21st Escort Group, I should like to remark upon the splendid cooperation I have received from all ships of the Group.
>
> I have the honor to be,
>
> Sir,
> Your obedient servant,
> [sig]
> Lieutenant Commander, D.S.C., R.N.[33]

The 21st Escort Group had set sail from Belfast at 5:00pm on April 21 with orders to proceed to Kyle of Lochalsh for exercises with the HMS *Philante*. This group consisted of six vessels in two divisions. The 1st Division consisted of HMS *Conn*, HMS *Rupert*, and HMS *Deane* and the 2nd Division HMS *Redmill*, HMS *Fitzroy*, and HMS *Byron*. The Group arrived at Lochalsh on the 22nd, carried out the exercises, and then departed that evening for Moville, at the mouth of Lough Foyle, arriving there the morning of the 24th. At Moville, the group topped off with fuel then loaded "High Tea" stores in three vessels: HMS *Conn*, HMS *Rupert*, and HMS Redmill. "High Tea" was the British code word for sonobuoys. The 21st Escort Group departed Moville on the morning of the 25th to relieve the 4th Escort Group on patrol in an area bounded by 55–56ºN and 08.45–11.15ºW. A U-boat was known to be in the area.

U-1305 (VIIC/41) under the command of Klt Helmuth Christiansen, one of the last of the VIIC/41 series to be commissioned, had departed Stavanger on its first war patrol on April 4. Its patrol area was to the west of the Hebrides. On April 24, a sound contact was made by *U-1305* and a T5 torpedo fired, striking and sinking the British steam merchant SS *Monmouth Coast*, northeast of Tory Island. RAF Coastal Command was now on alert and dropping sonobuoys while the 4th Escort Group was conducting a sweep.

As the 21st Group was en route it was alerted by Coastal Command aircraft to a "High Tea" contact. This contact could have been *U-1105* making its way south or *U-1305* heading west in the early morning hours of the 25th, but this will never be known with certainty. It should be noted that unlike the US Navy, which used *Ultra* information to vector in hunter-killer groups on a suspected U-boat, the Royal Navy took no such action, using *Ultra* in a more defensive role by rerouting convoys around areas where it was suspected U-boats were present. There was absolutely no warning issued to the 21st Escort Group that *U-1105* and other U-boats such as *U-1305* were operating in the area. The 4th Escort Group started the search, which was then taken over by the 21st Group.

Coastal Command aircraft marked another area, reporting that an oil slick was observed. However, this was a known wreck site that the 21st Group noted in their report was "a great favorite with aircraft." Coastal Command dropped depth charges on the possible contact as the 21st Group approached the area. Fitzroy obtained a contact and HMS *Byron* was ordered to join the sweep but the contact was determined to be a wreck split in two. The search continued through the night of the 25th into the morning of the 26th. Oil over the wreck sight was sampled and determined to have "a heavy character and bore no resemblance to diesel," confirming the original suspicion that this was a known wreck.

The 21st Group continued their westward sweep. At 4:00pm on the 26th HMS *Redmill* in the 2nd Division exchanged places with HMS *Deane* in the 1st Division, in order to keep the three ships carrying "High Tea" equipment together. The 2nd Division was then dispatched to patrol the area south of Little Minch, which is the northern end of the Sea of the Hebrides to the west of the Isle of Skye, while the 1st Division proceeded to patrol a near 100 nautical mile box that was bounded, generally, by the northern coast of Ireland up to the southern islands of South Uist, west just beyond the continental shelf. The 21st Group's patrol area was focused on U-boats approaching inland toward the Irish Sea through the North Passage.

The first sweep of the box was along the southern boundary of the patrol box, out to the 100m (110yd) line, and was completed at 7:25am on April 27. The 1st Division then turned northeast on a heading of 25° to sweep toward the location of the SS *Monmouth Coast* sinking. Two hours later along this course the HMS *Redmill* was struck by a torpedo in position 54.23°N 10.36°W while patrolling abreast with the other two ships of the division. HMS *Conn*'s ASDIC had picked up a possible contact that did not have the characteristics of a torpedo and was classified

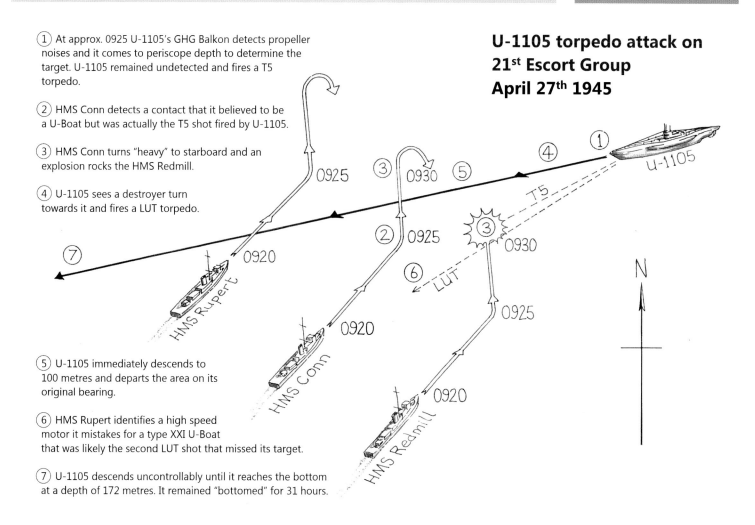

Figure 16. This depicts the most likely tactical situation of U-1105's only combat engagement as derived from careful analysis of both the 21st Escort Group's After Action Report and the personal recollections of Schwarz, U-1105's commanding officer. (Author's rendering)

as a "submarine". As we know from Schwarz's account, this was his initial T5 shot. As the contact faded off the starboard beam, and the *Conn* began a turn, "a heavy explosion occurred and a large plume of water was observed at HMS *Redmill*'s stern. It was immediately realized that it had been torpedoed. The original explosion was followed very shortly afterwards by a second, close on the port quarter, which threw up a plume of water several feet high. While Schwarz believed this to be a hit by his second Lut shot, this explosion could have been from an exploding depth charge. Later, several other explosions occurred near her stern which were undoubtedly some of her depth charges exploding deep." The HMS *Redmill* was ordered to look after itself as best as it could as the HMS *Conn* and HMS *Rupert* began a search for the U-boat. HMS *Rupert* picked

Figures 17 and 18. These photos were taken after the HMS *Redmill* was towed back to port where its damaged stern was inspected. The *Redmill* was deemed a total loss and scrapped. Only a single torpedo from *U-1105* likely hit the stern, with additional damage caused by exploding depth charges present on the *Redmill*'s deck launcher. (Courtesy of Maryland Historical Trust)

up what it believed to be a burst of high electric motors and initially thought it possible that the attack had been made by a Type XXI U-boat, but quickly altered that thinking. What HMS *Rupert* picked up may well have been the missed Lut shot.

The search for the U-boat proceeded southward in a 6 nautical mile square box toward the coast. It was believed that the U-boat "had retired to the south-westwards with an initial burst of high speed" where it was thought it would "take refuge on the bottom inshore where he could come to snorkel depth during the night to recharge." However, no firm contact was ever identified, though a suspected contact was attacked by a single Hedgehog salvo by both ships. The 2nd Division, meanwhile, was ordered to arrive on location to support. The search for the U-boat was soon called off due to a lack of contact and worsening weather that required HMS *Redmill* to be put under tow and escorted to port to prevent its sinking. Twenty-eight sailors on board were killed in the initial torpedo strike and resulting depth charge explosions.[34]

While the search was focused south-southeast, *U-1105* had dived, apparently on its original westward heading, bottoming unplanned due west of the immediate search box, where it remained for some 31 hours. The acoustic camouflage offered significant protection during the initial approach to the three destroyer escorts and subsequent evasive movement below them as *U-1105*'s direction of movement was not detected.

The historical record is clear, however, that there was no extensive search for *U-1105* due to the poor weather and the commitment to rescue HMS *Redmill* from sinking. While much has been made about *U-1105*'s Alberich being a key factor in its ability to avoid detection after hitting the HMS *Redmill*, its Alberich was not tuned for the depth at which it remained for 31 hours; it was tuned for a far shallower depth. While it is not clear what Schwarz might have heard, it does not appear that any Royal Navy vessels came close to *U-1105* in the subsequent search nor did they drop nearly 300 depth charges/Hedgehogs. A Royal Navy veteran on board one of the search vessels and the official historian of the 21st Group noted that "no depth charges were dropped because the *Redmill* had so many men in the water and their sonar could not locate the *U-1105*."[35] It is likely that *U-1105*'s very sensitive GHG Balkon amplified other sounds that were being misinterpreted as depth charges. This was the first patrol during which *U-1105* used this device and no doubt its operator had little experience in the interpretation of ocean sounds.

U-1105's acoustic camouflage at the depth at which the boat came to rest was aided by the thermoclines and bottom features. Alberich certainly helped avoid the U-boat's initial detection when it likely maneuvered past the Royal Navy frigates, but as documented in the previous chapter, "bottoming" could be just as effective in camouflaging a U-boat's signature. Schwarz confirmed this in a 1991 interview when he stated that in addition to Alberich, he attributed his boat's escape to "the fact that the U-boat had remained motionless on the bottom during the attack."[36] In this case, *U-1105* was lucky that it had not followed standard procedure and maneuvered close inshore to bottom, given that this was what the British had anticipated. That said, in shallower water *U-1105* would have gained the full advantage of its Alberich.

U-1105's patrol, as well as that of *U-1305*, proved how effective late-war U-boat innovations were. The snorkel allowed both U-boats to remain submerged for weeks, reach the English coast and operate there without detection. The GHG Balkon gave them a significant advantage in detection against all Allied vessels. Bottoming allowed the U-boats to hide along the coastline, maintain the crew's endurance, and allow them to keep on station in their assigned operational area. In the case of *U-1105*, its Alberich, tuned for shallow water, masked its approach against three

Figure 19. U-1105 is pictured on the surface cruising toward its surrender rendezvous point. This photo was likely taken by a British Sunderland aircraft that flew overhead on May 9, 1945. (Courtesy of Maryland Historical Trust)

veteran Royal Navy frigates as designed. Only the inexperience of the young German captains, both of whom were on their first war patrols and in their 20s, prevented them from being more effective in their approach and torpedo shots against their targets. Twelve months earlier this sort of performance by a U-boat was unthinkable by both Adm Dönitz and especially Allied intelligence. While many historians of the Battle of the Atlantic tend to focus on the convoy battles of the early war years or the triumph of the Allies over the U-boat's wolfpack tactics during the "Black May" defeat of 1943, almost none have grasped the evolution of the U-boat in the final year of war. This transformation in technology and tactics foreshadowed the extremely difficult and often dangerous hide-and-seek nature of submarine operations in the postwar world that are still relevant to this day.

On May 9, *U-1105* was cruising surfaced having sent its surrender signal that morning and given its position as 55.48ºN 11.16ºW. As instructed, it was flying a "Blue and White" flag. The crew was crowded outside on the conning tower. An RAF Sunderland aircraft on standby vectored to the position in order to escort the "Black Panther" to a rendezvous with a Royal Navy vessel. As the Sunderland approached it

Figure 20. *U-1305* is pictured flying the "Black Flag" from its attack periscope while surrendering under escort to the Royal Navy at Loch Eriboll on May 10, 1945. *U-1105* surrendered under similar circumstances this same day and at the same location. Note that *U-1305* had a Type II snorkel installation with below-deck air intake trunking on the port side. *U-1305* operated in the same area during its patrol as the "Black Panther". It was not coated with Alberich, but it managed to elude the Royal Navy after several unsuccessful attacks by the use of its snorkel and the tactic of bottoming. (Author's collection)

noted that the crew was "waving madly and giving the 'thumbs up'."[37] It was likely at this time that the Sunderland took two photographs of *U-1105* reproduced in this book.

U-1105's war diary was destroyed along with all of the other papers and messages before Schwarz surrendered his U-boat. The loss of the U-boat's log prevents there being a complete historic picture of *U-1105*'s final cruise, other than what has been shared by Schwarz. However, we do have the surviving war diary of *U-1305* as a guide. *U-1305* left Stavanger, Norway on April 5 and surrendered on May 8. Its cruise took it along almost the exact same route and patrol area as *U-1105*. *U-1305* was equipped with a snorkel and a GHG Balkon. It was also one of the last commissioned U-boats to conduct a wartime patrol. Its patrol lasted 34 days and during that time it never surfaced and remained submerged. It cruised for 1,699.2 nautical miles of which 818.3 were traveled while snorkeling and 880.9 while running on electric motors. It bottomed seven times and conducted one successful and three unsuccessful torpedo attacks on three separate days. While it was not equipped with Alberich, it was also never located by Allied patrol vessels. This was testament to its snorkel and tactic of bottoming. While we know that *U-1105* briefly surfaced

twice, the rest of its cruise likely followed a very similar pattern of snorkeling and electric motor employment as that of *U-1305*.³⁸

U-1105 arrived at the entrance to Loch Eriboll in the early evening on the following day, May 10. The U-boat was met by HMS *Conn*, which provided a boarding party. The HMS *Conn* was the very ship that had identified *U-1105*'s T5 shot and inadvertently turned toward the approaching *U-1105* on April 26. Neither the crew of the HMS *Conn* nor *U-1105* were aware of their deadly encounter two weeks earlier. Schwarz proceeded to maneuver *U-1105* into the Loch at 8:00pm. Schwarz wrote that:

> Loch Eriboll was reached on May 10th. There a British boarding party came on board. At their instructions, we and other submarines went through the North Minch to Loch Alsh, a British antisubmarine group base. That was where the first interrogation of the commandant by English naval officers occurred. In addition, most of the crew had to leave the submarine and transfer to a transport ship.
>
> With its remaining crew, *U-1105* then spent one of the next few days in a convoy with seven other submarines, escorted by frigates and aircraft through Little Minch to Lisahally near Londonderry, Northern Ireland.
>
> The Chief Engineer (Oberleutnant Ingenieur der Reserve Larsen)

Figure 21. *U-802* (IXC), unidentified (VIIC), *U-1105* (VIIC), *U-826* (VIIC) in Holy Loch shortly after surrender in early May, 1945. Note how dark and clean *U-1105*'s appearance is due to its Alberich coating when compared to the other U-boats docked alongside. Also of interest is the distinctive vertical water vents alongside its hull. (Author's collection)

remained behind with a speciallyselected crew of technical occupation members. He was tasked by the English with maintaining a "package" of five submarines. The remaining crew made the trip to become prisoners of war of the British.

U-1105 was later turned over to the USA as spoils of war.

On May 11, *U-1105* was taken under escort by the HMS *Rupert* from the 21st Escort Group in the early morning and it departed for Loch Alsh, near Skye in northwest Scotland. Loch Alsh had immediate importance.

Loch Alsh had been designated as a "Port for Final Examination", the principal roles of which were the removal of the U-boats' torpedoes, as well as taking the majority of the German crews into captivity. Thus *U-1105* had to go through this process before it could be moved to a "Laying-up Port" to await decisions about its ultimate future, in this case Lisahally in Northern Ireland.[39]

Thus began *U-1105*'s postwar career, which exceeded its wartime role by more than half a century. *U-1105* participated in extensive postwar testing and evaluation. It was also the last German U-boat to cross the North Atlantic.

CHAPTER 4

POSTWAR EVALUATION AND TESTING BY THE ROYAL NAVY

It is through the postwar testing period that *U-1105*'s value emerges not simply as a war prize, but also as a means to evaluate the two key technological developments during the war – the snorkel and acoustic camouflage – that drove the future evolution of submarine warfare.

The first discussion of postwar testing of the surviving U-boats came in the form of an Admiralty meeting on June 25, 1945. The meeting notes stated that U-boats held at Lisahally and Loch Ryan in southwest Scotland were: "manned on the basis of 1 German and 1 British crew to 5 submarines, employing in all 400–500 British officers and men. The USA has a small group at Lisahally but the British are acting as 'caretakers' of the U-boats for the United Nations and will continue to do so during the trials which have been authorized by the Admiralty: these trials to be carried out in conjunction with the USA authorities."[40]

The Allies had access to more than 125 U-boats that either surrendered or had been abandoned in German and foreign ports. Despite the fact

that the U-boats in the UK had not yet been reviewed by the Allies and that no decisions had been taken concerning their futures, the Royal Navy made a unilateral decision to subject a few of them to extensive evaluation. The trials were focused primarily on testing the Electro-boats in the form of a Type XXI (either *U-2502* or *U-3017*) and a Type XXIII (*U-2326*) with British crews for seven days of general testing, followed by 30 days of "fast target" antisubmarine trials. Additionally, it was decided to conduct comparative testing trials between two Type VIICs: one with "rubber", which was *U-1105*, and one without, which was *U-1171*. *U-1171* was christened the "White Puma" by its Royal Navy crew as it as painted white in order to be able to tell it apart from the "Black Panther" during trials.

The proposed testing program of these two U-boats consisted of antisubmarine trials for 14–15 days, including radar testing of the rubberized hull; deep-diving detection trials for five days to include high-frequency direction finding; antisnorkel trials against the Type XXI and Type XXIII U-boats; aerial bomb sight trials against a Type XXI snorkel; sonobuoy trials; antiradar trials on VIIC snorkels with and without rubber covering (this is additional confirmation that *U-1105*, did not have a Wesch matt-coated snorkel); noise trials in Loch Goil with the Type XXI and XXIII U-boats and with *U-1105*; high-speed noise trials at speeds

Figure 22. *U-1105* at Holy Loch docked next to *U-826* (off picture to the right) on May 14, 1945 shortly after its surrender. This close-up shows what appears to be the moment that Royal Navy officers took command of *U-1105* as evidenced by the two rows of former enemies standing opposite each other while at 'parade rest' on the conning tower. (Author's collection)

Figure 23. *U-1105* at Holy Loch docked next to *U-826* (right) on May 14, 1945, shortly after its surrender. This picture shows both Royal Navy crews conversing with each other from their respective conning towers while the German crew of *U-1105* prepares for departure. (Author's collection)

Figures 24 and 25. Two images of *U-1105* as it maneuvers in Holy Loch, possibly after a trial. (Courtesy of the Royal Navy Submarine Museum)

considered "safe and shallow"; thermal detection on the Type XXI, infrared detection on Type XXI snorkel coated with anti-infrared paint; torpedo discharge trials, battery charging, ventilation and hydrogen content trials, electrical maneuvering; and trials against different snorkel heads. These varied goals demonstrate just how important the snorkel, as well as both acoustic and antiradar camouflage, was. What was interesting about the testing of the "rubber"-covered boats in the meeting notes titled "Trials to be Carried out with Surrendered U-Boats" issued on June 26 was the recommendation that both sonar and radar tests were to be carried out as the British were still unsure of the exact purpose of Alberich.[41] In order to

Figure 26. *U-1105* during testing trials with the Royal Navy in Holy Loch. Note the starboard side exhaust trunking and shut-off valve that runs along the forward section of the conning tower. This remains a prominent feature on the wreck to this day. On the port side the snorkel mast clamp extends past the conning tower and just below is the flange valve of the snorkel intake. When the snorkel mast was raised, another flange of the snorkel mast would create a watertight seal and allow fresh air to be sucked into the U-boat while it was underwater. The running light is mounted on the attack scope mast indicating that the U-boat was likely not being used for tactical maneuvers at the time the picture was taken. Also note the watertight tube that extends aft along the starboard side upper Wintergarten. This feature was unique to *U-1105* and is a prominent feature on the wreck today. (Courtesy of the Royal Navy Submarine Museum)

Figure 27. *U-1105* possibly preparing for one of its numerous tests in Loch Goil while commissioned with the Royal Navy. Note that the British maintained its unique conning tower emblem of a "Black Panther" sitting astride the North Atlantic. The image was taken from the deck of *U-1171*. (Courtesy of the Royal Navy Submarine Museum)

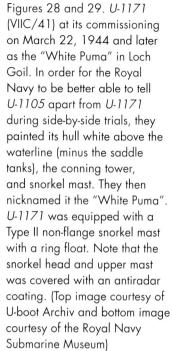

Figures 28 and 29. *U-1171* (VIIC/41) at its commissioning on March 22, 1944 and later as the "White Puma" in Loch Goil. In order for the Royal Navy to be better able to tell *U-1105* apart from *U-1171* during side-by-side trials, they painted its hull white above the waterline (minus the saddle tanks), the conning tower, and snorkel mast. They then nicknamed it the "White Puma". *U-1171* was equipped with a Type II non-flange snorkel mast with a ring float. Note that the snorkel head and upper mast was covered with an antiradar coating. (Top image courtesy of U-boot Archiv and bottom image courtesy of the Royal Navy Submarine Museum)

help determine the rubber coating's purpose once and for all without compromising *U-1105*'s Alberich, samples from the surrendered *U-485* were removed and sent to the US Naval Research Lab, which later confirmed Alberich was for antisonar purposes.

The testing program for the two Type VIICs had priority. While neither U-boat was formally commissioned into the Royal Navy, both were given an N-series Pennant Number. *U-1105* received Pennant Number N.16 and *U-1171* received Pennant Number N.19. Both numbers were painted on their hulls and they transferred from Lisahally to the 3rd Submarine Flotilla (HMS *Forth*) in Holy Loch, arriving in western Scotland on June 29.[42]

Inevitably, the extensive U-boat testing program proposed in June ran into logistical and political realities. In a memo dated July 27, it was noted that the trials of *U-1105* and *U-1171* originally drawn up by the Admiralty could not be completed and that any proposed dates should be viewed as a "guide."[43]

The month of August brought the first trials of *U-1105*. German experts were on hand for these tests. On the 6th of the month *U-1105* was moved from Lisahally to Coll Sanctuary, where it began sonar testing with *U-1171* and HMS *Kingfisher*. *U-1105* was dived twice during the tests: the first submerged trial lasted for three hours and the second one lasted for one hour. Another test on the 7th introduced an aircraft search, probably using radar against Alberich when the U-boat was surfaced. On the 8th, testing began against the snorkel. Two snorkel runs took place, one lasting approximately 45 minutes and the second lasting 30 minutes. This series of tests was completed on the 11th and *U-1105* was moved back to Lisahally.[44]

At the start of the latter snorkel testing the Captain (Submarines) of HMS *Forth* reported on August 8 that:

"Both [Type VIICs] have completed diving, deep diving, full power and snorkel running trails without any serious hitch. One of these U-boats [*U-1105*] is completely rubber covered, and the purpose of the forthcoming trials is to discover the effect of this as an anti A/S measure. The immediate problem is to re-stick various areas of rubber which has begun to peel off, and the attention of several scientists for nine days in dock has failed to achieve much."[45]

While this was noted, trials continued.

Additional requirements were issued by the Air Ministry to the Admiralty on August 13 for comparative tests between *U-1105* and *U-1171*.[46] On the 14th, those tests were attempted but cancelled due to poor surface visibility. The tests were later conducted during the 15th and 16th at Lisahally. However,

OPPOSITE Figures 30 and 31. *U-1105* docked next to *U-1171* in Holy Loch before *U-1171* was painted white for trials. Both images provide excellent views of the conning tower and Wintergarten area visible on the wreck site today. *U-1105* is at the bottom of both images. The left image shows a superb view of the external exhaust trunking. The submarine to the far left/right is the HMS *Tuna* (N94). (Author's collection)

Figures 32 and 33. *U-1105* during Royal Navy trials in Loch Goil. (Courtesy of the Royal Navy Submarine Museum)

technical issues were experienced with the water circulation system on the 17th that required German crew assistance to fix.[47]

On the 19th, *U-1105* departed Lisahally for Holy Loch. The U-boat entered a floating dry-dock on the 21st where work began on scrapping the hull to repair the sections where the Alberich was damaged or worn, as well as painting. Alberich sheets were removed from the Alberich-covered *U-485*, which were taken from the same relative positions on the boat. On the 24th, work on repairing the Alberich was compete and the was dry-dock flooded. *U-1105* departed Holy Loch on the 25th and cruised to Inchmarnock (north) and Rothesay where testing commenced with a deep dive, then a snorkel run. Snorkel testing continued on the 26th, followed by a move back to Holy Loch. From the 27th until the 31st, *U-1105* conducted side-by-side testing with *U-1171* at Loch Goil and Lisahally.[48]

At the end of August an evaluation of the overall state of *U-1105*'s readiness was conducted with a report being issued on September 4. The report read:

Report on the state of U-1105

Date of last docking: 21st August, 1945
Hull: Good condition. Rubber covering, however, is in a bad state.
Main Engine: Krupps, Germania Werk. Non-reversing. Satisfactory condition. All valves require overhaul, before full power can be achieved. Muffler valves leak badly under diving pressure. Several attempts have been made to remedy this, but it appears to be an inherent weakness in design in this type of U-boat.
Propeller and Shafting: Both good. Thrusts, brakes, stern glands are all in good condition.

Auxiliary Machinery: In general the auxiliary machinery is in a satisfactory condition with the following exceptions:
Electrical Air Compressor and Junkers Air Compressor: In working order, but both require complete overhaul.
Forward Telemotor Pump: Noisy in operation and require complete overhaul.
Electrical Lub. Oil Pump. Noisy in operation, requires overhaul.
Main Motors: In good condition.
Main Battery: In good condition, date of installation unknown.
Main and Auxiliary Switchboard: In satisfactory condition.
Miscellaneous:
W/T Machinery. In good condition.
Echo Sounder. Defective. Zero insulation between oscillator and amplifier. Control box burnt out.
Log. Defective.
Hydrophones. All gear in good condition.
Radar. Satisfactory condition.
GSR. Borkum: Satisfactory condition. Naxos: 3cms and 9cms lack Tunis aerial.
Torpedo Tubes. Satisfactory condition.
Spares. Approximately 30% have been used. Remainders are in a satisfactory condition.[49]

U-1105 remained at Holy Loch at the start of the month. A snorkel run for another trial was conducted on the 4th at Loch Goil. Cleaning of the

Figure 34. *U-1105* during trials in Loch Goil. Note the aft navigation light on the portside, aft-quarter that can be seen slightly raised above the deck. On the upper Wintergarten a white horseshoe life preserver can be seen. This was added by the British crew and had the name "Black Panther" spelled out in English. (Courtesy of the Royal Navy Submarine Museum)

U-boat occurred on the 9th, apparently in preparation for the Tripartite Naval Commission (TNC) inspection that commenced at 9:00am on the 11th. That same night at around 9:45pm *U-1105* left its berth, dived and began a snorkel run to Loch Goil, completing the transfer at 11:15pm. The next day, *U-1105* conducted additional trials with its snorkel then returned back to Holy Loch. Submerged trials and a snorkel run again occurred at Loch Goil on the 14th. Two days later, on the 16th, *U-1105* moved to Fairley and Rornesey where additional tests occurred until the 17th before it returned to Holy Loch. The testing from the 4th–17th included thermal radiation detection of the snorkel followed by noise trials at Holy Loch on the 18th, Fairlie on the 19th, Rosneath on the 20th, and Fairlie again on the 25th.[50]

U-1105 started the month of October at Holy Loch. A report noted that it was "in good condition" and was maintained by a German crew under British supervision.[51] It moved to Holyhead on the 1st and remained

there until the 19th, and various trials were conducted with *U-1171*. While in Great Ormes Head area on the 9th the starboard engine clutch failed and had to be stripped and repaired. *U-1171* stood by, presumably to assist. Testing recommenced along with a snorkel run on the 11th.[52]

An inquiry was received by the Admiralty from the US Navy as to when *U-1105* would be made available to them for testing, as it had by then been formally allocated to the USA by the TNC in Potsdam. An internal response back to the Admiralty from the testing authority was issued on October 11th and stated:

To Admiralty

DHND 091510 Para. 1B
1. *U-1105* which is rubber coated, is manned by a British crew and is at present carrying out special comparative trials for H.Q.C.C. D.A.W.T D.N.A.R. D.R.E. and SSE (H)
2. It is expected that these trials will complete early in December.
3. Propose US Naval Authority be asked if *U-1105* can be retained until completion of these trials, observing that all data obtained from these trials will be made available to them.[53]

The Admiralty sent a response to the US Navy on the 16th asking permission to postpone the delivery of *U-1105* until all trials had been concluded.[54] The US Navy responded in the affirmative on the 23rd that the Admiralty could retain *U-1105* until the completion of its testing.[55]

Despite this seemingly supportive exchange that *U-1105* had been allocated to the USA, there was some confusion as to who actually would retain *U-1105*. It appears that the Senior American Representative to the TNC, ADM Robert Ghormley, incorrectly advised the TNC that *U-1105* was not required, along with several Type XXI and Type IIIs that had previously been allocated. At the next TNC meeting toward the end of October the Soviet representative expressed interest in acquiring *U-1105*, given the perceived US position. ADM Ghormley sent a message back to the US Navy's Chief Naval Operations on October 30 asking for clarification on the current position regarding *U-1105*. He stated that this U-boat was experimental and covered with

Figure 35. *U-1105* in dry dock at Fort Blockhouse, England, November 1945. The wear of constant trials is evident by the peeling Alberich off the rear quarter of the starboard side. (Courtesy of Maryland Historical Trust)

rubber for antisonar purposes. While that message was being reviewed in Washington, DC, the Soviet TNC Senior Representative Admiral Gordei Levchenko asked the TNC to receive both the *U-1105* and the *U-1305*, the latter being among the last U-boats commissioned during the war. The Soviets had expressed an early interest in Alberich, even more so than the Walter turbine-equipped *U-1407*.

During the course of several days in late August and early September a team of British, Soviet, and Americans conducted the first joint inspection of surrendered German U-boats in Lisahally, Loch Ryan, and Barrow as part of the Tripartite Commission. The inspection was conducted in part to determine the final allocation of U-boats to each of the three countries. The Russians took an interest in both the Alberich and Wesch matting of

Figure 36. *U-1105*'s starboard side bow area just behind the torpedo tubes while in dry dock and Fort Blockhouse, November 1945. The two large round circles are *U-1105*'s starboard side UT microphones and the nine smaller circles are the original GHG microphone array. The Alberich is peeling along the lower bow. (Courtesy of the Royal Navy Submarine Museum).

U-485 in Loch Ryan. As noted in the report:

> There was one incident at Loch Ryan – our friend Bondarivk asked if he could take samples of the rubber covering one of the submarines there. He was told "no" by Captain S/M, who quite rightly quoted my instructions that nothing was to be removed from the U-boats by any member of the Commission. Later he was seen to put a piece in his pocket, and, also against orders, to remove a piece of the antiradar composition from a *schnorkel*. On this fact being reported, I saw Sheskayev and Orel and told them I was pained to hear of his conduct, not that the articles in question were important in themselves, but on principle. This shook them and they both apologised and dropped

down on Master Bondarivk pretty severely. I said the samples were to be returned – which was done.[56]

The rubber coating resonated with the Soviets and clearly influenced their desire to obtain *U-1105*. The issue about who would retain *U-1105* was concluded on November 1 when the US Navy reversed its earlier decision and communicated to the TNC the fact that it wanted to retain *U-1105* and not *U-805*.[57] This decision, however, is perplexing after more than half a century, as the US Navy never once conducted any operational trials relating to *U-1105*'s snorkel or Alberich once it arrived in US waters.

On October 12, both *U-1105* and *U-1171* were docked together at Beaumaris. Extensive trials continued through the 20th.[58] On October 22, *U-1105* put in at Fort Blockhouse, Gosport for an overhaul as it had developed "serious defects" since the inspection by the TNC team in August as a result of all the extensive testing. A report from Flag Officer (Submarines) noted the following defects:

- Both engine clutched [sic] require striping [sic] and refitting
- Both superchargers were defective and required refitting
- All exhaust boxes, injectors and starboard air start valves require refit
- Port and starboard mufflers and group exhaust valves require stripping and refit
- Rubber covering peeling badly

Trials can probably be completed on main motors by 3rd November.

The complete retrofit of the diesel engines while at Fort Blockhouse was anticipated to take four weeks from November 3. Once the retrofit had been completed it was thought that *U-1105* would be ready for the transatlantic crossing under its own power "but rubber coating if renewed is certain to deteriorate considerably on passage."[59] The report recommended that the US Navy send its own inspection team to look over *U-1105* and independently assess the work that needed to be completed. One thing was clear, the Royal Navy was keen to continue

Figure 37. The author is pictured holding a 5 x 45 x 45cm (2 x 18 x 18in.) piece of Soviet-produced anechoic tile for scale. The Soviets aggressively exploited German U-boat technology after the war. They were particularly interested in Alberich and began to coat their submarines with what is known today as anechoic tiles some 25–30 years before the US Navy. This piece of tile came from the Soviet Project 651 (K-77) (NATO designation Juliette Class 484) commissioned on March 11, 1965. The picture on the right illustrates how the Soviet engineers utilized the same basic principal of holes to help absorb sonar sound waves, just like on *U-1105*.

working *U-1105* as hard as possible and eventually conducted a final test with *U-1171* on November 31.[60]

The US Navy deployed an inspection team to Fort Blockhouse to look over *U-1105*. They sent an official response to the Admiralty on November 8 instructing them to commence on the recommended refitting required to make *U-1105* seaworthy for a transatlantic crossing except for the repair of its "rubber covering which, due to voyage damage, would probably result in wasted effort."[61]

U-1105 was made ready to sail on December 12 to the USA. It was the recommendation of the Admiralty that *U-1105* sail to the USA through a southerly route that included possible stops at the Azores and Bermuda for refueling if necessary.[62] A southern route would presumably take *U-1105* through calmer Atlantic waters and afford it the potential to make stops if mechanical issues were experienced. The Admiralty had put *U-1105* through extensive testing, and despite the overhaul, this was a German U-boat and not a British submarine. The British technicians could only fix so much given their lack of familiarity and specific spare parts, even with the technical supervision of former German crewmen who were retained to assist during the trials.

The three months of Admiralty trials resulted in a number of conclusions. Detection of German snorkel exhaust by thermal radiation resulted in the following assessment:

> It can be concluded that, using lead sulphide cells with their present sensitivities and used in conjunction with very large optical systems detection of a Schnorkeling submarine by radiation from exhaust gases is impossible at ranges of 200 yards or greater. Similar considerations apply to thermocouples at ranges of 400 yards.[63]

During September's noise trials conducted between *U-1105* and *U-1171*, as well as with *U-2326*, It was determined that the Alberich had no effect on the transmission of sound from machinery to the water. It was noted that, while the *U-1105*'s main motors were slightly nosier than *U-1171*'s, they were both "very quiet machines." It was determined that the German U-boat was generally significantly quieter than British submarines, primarily due to the slower revolutions of a U-boat propeller. *U-2326* proved "quite inaudible up to about four knots when using the flexibility mounted 'creep' motor." When it used the main motor drive through the gearbox then its sound became prominent. The use of a snorkel at normal operating speed and the main engine produced "very considerable increase of noise over that obtained with main motor and gear box drive."[64] A December report issued on the testing of variable field effect demonstrated no appreciable advantage for *U-1105* over *U-1171*.[65]

While no specific report has been located that provides the result of any Alberich testing against *U-1105*, it was known from German testing that it could reduce sonar reflectivity by 20 percent, thereby reducing the ASDIC detection range from anywhere between 20–60 percent. Its effectiveness was based on the depth to which it was tuned and where in the water column the U-boat was operating at the time of the ASDIC ping.

British sailor John Edge Woodcock, 3rd Class ERA, was a crewman on board *U-1105* during its testing. Below are his postwar recollections of that time:

> After ferrying a couple of U-boats from [Stavanger] back to Londonderry, I was allocated to an older Type VIIC U-boat, *U-1105*.

The "Black Panther" was fully encased in rubber, even the armament around the gun, etc. It was supposed to deaden the "ping" I believe. Anyway this U-boat and another of the same type with no rubber covering were commissioned. After a lot of pipe tracing, checking and re-checking, not helped either by the language difference and the nonexistence of English/German dictionaries, we managed to find out what things were for. . . .

The captain was a Lt Cdr RN. I was Chief ERA of the boat. Roy Wilks was the outside ERA.

Within a few days the U-boat was taken out, trimmed & dived, etc, and all manner of things were done, Schnorkel running which I must say kept the engine room quite busy and really on their toes. This type of running was OK if the sea was not too rough and had good hydroplane control of depth and angle etc., otherwise it caused a hell of a vacuum in the boat, not noticed too much in the engine room as air intakes from the snorkel come into each side of the engines around the hull. In the forward part of the boat it is a lot different, a very slight vacuum is felt at a very early stage and if the depth is not controlled fairly quickly they get very short of air to breath [sic].

We did a lot of exercises with surface craft and Coastal Command in various phases. At one time we were in a Loch in Scotland and were dived between two buoys and had to run the engines for hours on the Schnorkel, so they could tryout their Infra-Red detection on the shore. I believe it was highly secretive in those days.

These U-boats were not built for comfort, they were strictly a war machine and a very good boat too, very well built and were able to withstand all weathers of which I will relate fully towards the end of my story.[66]

At the end of November the transition of *U-1105* to the US Navy was underway.

CHAPTER 5

NORTH ATLANTIC TRANSIT TO THE USA

It was implied in the Maryland Historical Trust's National Register of Historic Places Application for *U-1105* that the US Navy required *U-1105* for testing because of its acoustic camouflage – Alberich.[67] This suggestion is not confirmed by any period documents discovered to date. If the Alberich sheets were of importance, the US Navy could have simply stripped the sheets off the hull in England and sent them back to the labs in the USA for testing without bringing the entire U-boat to the USA. We know that US technical inspection teams had already examined the Alberich-covered *U-485*. Strips of the Alberich could have been removed at that point but were not.

The US Navy's Technical Mission in Europe had already reached German production facilities on the north coast in May 1945 and sent back to the USA an extensive list of materials used in the production of acoustic camouflage under what the Bureau of Ships labeled Project Code 332 and 470 respectively.

The following items were listed as being sent to the Bureau of Ships under Project Code 332:

1. Six (6) samples of soft rubber sheets with various type of perforations including one sample showing type of lining used inside fuel and water tanks.
2. Sample of abrasive paper used for grinding rubber sheets to thickness.
3. Three (3) bottles of cement.
4. Two (2) glass plates with rubber sheets cemented to them to show method of bonding and effects of cementing edges together.
5. Sample of rubber sheet as it comes from vulkanizer.
6. Sample of rubber sheet after surfaces have been ground.
7. Sample of rubberized cloth used to attach sheets together before perforating.
8. Sample composed of six (6) small pieces cemented together at edges.
9. Five (5) cutting tools used in original method of making perforations.

The following samples of adhesives and assorted materials were sent to the Naval Research Lab (NRL) under Project Code 470:

1. Sample of Buna S DEFO 1000.
2. Sample of Russ-Elastic.
3. Sample of Russ P 1250.
4. Sample of Schwefel.
5. Sample of Zinkweiss.
6. Sample of Vulkazit – D - M.
7. Sample of Vulk D.
8. Sample of Harz Amerik.
9. Sample of Aktiplast.
10. Sample of Kautshal.
11. Sample of Erweicher P l E.
12. Roll of non-perforated rubber 1 meter wide.

The NRL received from Bremen a triangular piece of perforated acoustically soft lining material that was mistakenly identified as Alberich at the time of shipment, but turned out not to be upon arrival at the lab.[68] All of this material was tested and evaluated by September 1945 when a formal report on acoustic camouflage was issued by the US Navy, three months before *U-1105* was to cross the North Atlantic. The report was titled "US Navy Technical Mission in Europe Report No. 352-45, Rubber Covering of German Submarines Anti-Asdic (German Code Name 'Alberich')". One copy of the report was ordered to be sent to the Bureau of Ships Section 940/330C to be placed in File Sonar Serial 3936 (940a) "for future study and reference." Two additional copies were sent to the Massachusetts Institute of Technology (MIT) "for use in connection with the currently assigned development problem. . ." All of the British testing reports about any exploited U-boat technology were also made available to the US Navy months before *U-1105* was to cross. It should be noted that the NRL's Director at that time was CDR Henry A. Schade, USN, who in 1945 had headed the US Navy's Technical Mission in Europe that had a lead role in the search and acquisition of the latest German U-boat technology across northwest German ports. He was one of the very few US Navy officers who had first-hand experience with late-war German U-boat technology.

The only conclusion that can be reached about *U-1105*'s transfer to the US Navy and transatlantic voyage was that it was intended to keep this particular U-boat out of Soviet hands. It certainly did not undergo any operational testing against its acoustic camouflage while in US Navy control. In fact, the US Navy did not conduct a single operational test of *U-1105*. The "Black Panther" was immediately assigned to detonation testing. The fact that it was finally sunk within the confines of an inland waterway and not in the open ocean also appears to corroborate this assertion, though no document located to date specifies a reason for the transfer.

One final point should be made. While the Soviet Union did in fact continue German wartime research into Alberich and began coating their submarines accordingly in the 1950s, the US Navy ignored this technology. Not until 1980 – almost 40 years after Alberich had been operationalized by the Kriegsmarine – did a US Navy submarine launch with antisonar anechoic tiles.

Figure 38. US Navy LCDR Hubert T. Murphy pictured here receiving the Bronze Star on the deck of the USS *Snapper* after completing six Pacific war patrols. Despite his wartime service, Murphy recalled later in life that it was his command of *U-1105* that stood out as the highlight of his military career. (Courtesy of Janet Murphy)

The US Navy dispatched a small crew to England commanded by LCDR Hubert (Hugh) Murphy, a highly decorated and respected US submarine commander from the Pacific Theatre. Murphy was the former Executive Officer on the USS *Greenling* and Captain of the USS *Snapper*. In a 2005 video interview conducted on Veteran's Day Murphy related how he was selected for this particular assignment. "In October of '45 we were sent to the Boston Navy Yard [with the *Snapper*] for decommissioning. While we were there, there was a need to pick up a German U-boat in Portsmouth, England and take her back to the US. Since I was the only bachelor skipper [available] at that time, in that location, and it was close to the Thanksgiving and Christmas holidays, I was chosen by the Squadron Cmdr to be the lucky guy that would take the ex-*U-1105* back to the states."[69] With a prize crew of 38, Murphy boarded the Aircraft Carrier Enterprise (CV-6) one early November morning and arrived at South Hampton six days later after an uneventful Atlantic crossing.

Despite the coincidental circumstances that brought Murphy into command of *U-1105*, he was undoubtedly one of the best officers to be chosen for this assignment. His US Navy efficiency reports during his time on the *Greenling* and *Snapper* reveal Murphy to possess a calm demeanor, and he was cool under pressure and well respected by his fellow officers and sailors alike. His superiors thought highly of him, noting after his wartime command of the *Snapper* that "Throughout the period of this report he has demonstrated an exceptional ability to command."[70] Murphy was awarded the Bronze Star for his final war patrol against Japan.

Murphy inspected *U-1105* on December 6 and 7, along with his First Officer and Chief Engineer. The U-boat was not in good shape after extensive testing by the British. This level of activity was not accompanied by the normal maintenance that a boat of this type might receive in German service. After the inspection a discussion occurred as to why it was so important to bring this U-boat back to the USA under its own power in a winter crossing of the North Atlantic. Murphy and the prize crew had not been given any information before they departed. According to William Ferguson, who was a member of Murphy's prize crew:

Figures 39 and 40. Two images of *U-1105* taken by LCDR Murphy before it departed for the USA. To the left of *U-1105* are the Royal Navy submarine HMS *Tuna* (N94), and the tug HMS *Earner* (W143). Note the horseshoe life preserver with the name "Black Panther" fashioned by the Royal Navy crew. It was apparently affixed to the conning tower in the photo on the left. Also of note is the flag pole holder shown to the right of the life preserver on the upper Wintergarten. This was likely stored in the watertight container affixed to the starboard side of the Wintergarten – a feature unique to *U-1105*.

> ... we all talked about why it was so necessary to get this one back to the states. We finally agreed on a preliminary list – the boat was built in 1943 and had snorkeling equipment for charging batteries while submerged at approximately periscope depth. It was completely covered with rubber coating to help escape our sonar and their periscope and optical equipment [was] better in some ways than ours. Their batteries could go longer without charging and required less watering.[71]

While this rationale was sound, the US Navy already had a half-dozen captured snorkel-equipped U-boats in US ports. With the exception of the Alberich, there was nothing unique required by the US Navy. Several surrendered U-boats in US Navy custody were also equipped with the GHG Balkon. As already pointed out above, the US Navy already had acquired every component required to test all phases of Alberich production. Even Murphy conceded in his postwar video that while the *U-1105* "was entirely rubber covered . . . in an attempt to make her more or less of a prime target for sonar . . . it turned out that it wasn't that effective, and as a matter of fact they said it was hardly noticeable, but it was part of the agreement that we would take her after the British had done the sonar experiments." One must ask the question as to why *U-1105* was to carry out a transatlantic crossing during winter only to be allocated to detonation testing upon arrival to the USA. No document uncovered to date has shed any meaningful reason for the transfer of *U-1105* from England to the USA. This arguably remains one of the last mysteries of the "Black Panther" that may never be solved.

Murphy and his prize crew conducted several familiarization cruises with *U-1105* in order to learn how to operate the U-boat. According to Ferguson, who was assigned to operate the "main propulsion", the crew had orders not to dive as "it was rumored that there may have been detonators set for depth or in the air lines placed by the Germans for scuttling purposes." This rumor appears self-imposed. As discussed above, *U-1105* was dived multiple times extensively – and deep – by the British during months of testing. In addition, all scuttling charges would have been removed before the testing began. The source of this rumor or its purpose remains a mystery.

Spare parts requested by the US Navy were soon loaded and final repairs to the U-boat completed. *U-1105* was expected to be ready by the 13th for its departure to the USA.⁷² Murphy requested through his Chief Engineer if several British ratings who knew the U-boat well would come along. Among the British who agreed was Woodcock.

U-1105 departed Portsmouth, England on December 19 and arrived at Portsmouth, New Hampshire on January 2. This was the last WWII German U-boat to cross the Atlantic. Woodcock's recollection of the crossing follows:

> We were soon on the way to Portsmouth & Dolphin. Several odd jobs were done with the help of spare crew members. At this time there were a few rumours flying about. One was that we were going to take the U-boat to the USA, but this was soon squashed as the Americans had arrived to take it over themselves. After a few trips with us and plenty of nattering, it was handed over to them. It was noticeable that they had people who could speak German and others had English/German dictionaries.
>
> My opposite number, their Chief [LCDR Murphy], was not too confident and he asked me if I would like to go across with them. I agreed to this if my outside ERA Roy Wilks could come too. This was then agreed, and the U-boat was stored up, etc, in readiness for our Atlantic trip.
>
> The American crew then gave the U-boat a good clean up for the trip. They were so eager to go as they hoped to arrive at Portsmouth, New Hampshire before their big day on 1 January 1946. After several delays through gales we eventually left about 19 December. There was [sic] still plenty of rough seas about but they decided they had to go.
>
> We had been out about four days when the problems started, it was so rough that we had to close the conning tower hatch and use the Schnorkel air intake, but with the mast still on deck. Quite a lot of water was coming into the engine room with the air and that caused the bilge to fill up rather quickly. This was OK when the pumps were clear and free of obstructions i.e. cotton waste, etc, which finally did the damage, blocking up the strainers, etc, and water was coming up to the deck plates. We eventually cleared the problem and away again.

I only hope they got the message. It is not a good idea to use cotton waste in the engine room for cleaning purposes.

Ferguson offers a slightly different account:

We pulled out of Portsmouth Harbor to cross the North Atlantic at the height of the storm season on the surface – December 19, 1945. After about 2 days out we started into very heavy seas and by the fourth day were in the middle of a hurricane we lost radio contact with England and then had to leave the conning tower hatch open – to be used as air intake for our engines keeping the main induction closed. We worked our way slowly in the severe storm taking water through the conning tower's open hatch and pumping the bilges continually. I seemed to be continually clearing the bilge pump strainer and replacing blown fuses.

Both Woodcock and Ferguson's accounts are in agreement over the severity of the storm, water intake, and issues with the bilge pumps. They differ on the use of the snorkel's employment. It appears that Woodcock's account is likely more accurate in that the snorkel was employed after the conning tower hatches were closed. The US Navy crew possessed only a cursory understanding of the snorkel, while the British crew had operated it multiple times over the last three months. This was important given that one diesel was temporarily lost and the U-boat was hit so hard by waves that it experienced two 80-degree rolls that threatened to turn the *U-1105* over. Had the conning tower hatch remained open, no doubt that seas could have easily swamped the U-boat and sent it to the bottom. It appears likely that the crew closed the hatches, raised the snorkel mast, and engaged the system without *U-1105* ever submerging in order to take advantage of the air intake valve's higher elevation above the heavy seas. Since the subsequent damage report recorded that the snorkel mast was bent, this suggests that the snorkel mast had been fully raised during the stormy part of the crossing and struck by a strong wave. Woodcock's account continues:

Our next problem was that a fire had started behind the starboard engine on the exhaust trunking caused by a fuel oil pipe joint leaking and running on to the trunking and eventually catching fire. When my attention was called to the fire I found the engines still running giving the fire plenty of air to boost up the flames. The engines were stopped, fuel shut off and the fire put out thanks to Roy Wilks and his quick thinking.

We then went on our way again, and by heavens it was rough. Watches were being kept on the periscope as they could not possibly stay on the bridge. We had not been getting much hot food either because of this bad weather but we kept pressing on.

The next little episode was again in the engine room, I was asleep in my bunk when I heard someone call the Chief and say something about one of the port engine exhaust valve top securing nuts. First of all it did not quite click, then I dashed to the engine room, but alas too late. They had put air start on that engine to start. The exhaust valve went down with the piston and then jammed under the cylinder head, bending valve push rods, etc, in the bargain. We then had to replace damaged parts which made it more awkward as the starboard engine was still running and with it being very rough you had to jamb your backside on to that engine for balance. This work was being done by Roy Wilks and myself. I am afraid the Americans were not used to this sort of work and conditions.

We finally boxed everything up and went on our way again with only one other slight snag, the engine clutches were slipping. They were friction type and we managed to adjust these and they held OK for the rest of the trip. As we reached the other side the weather abated, and then normal watches were kept on the bridge. We arrived at our destination Portsmouth, New Hampshire, on 2 January 1946 after about 14 rough days and nights.[73]

Ferguson recalled that:

Finally, after being in the storm for 10 days we had to shut down one of our engines; it had thrown a bearing. We tried our best to repair it and did finally, but only temporarily and could use it sparingly to maneuver. Our Quartermaster was finally able to get his sextant set up and after a while establish our position.

U-1105 found itself off Newfoundland near the 200m (218yd) line as many U-boats did in prior crossings. It had become the last U-boat to cross the North Atlantic and the only snorkel-equipped Type VIIC to do so.

Now out of the storm, Murphy set a southwest course for Portsmouth, NH. During the crossing *U-1105* lost all communications for ten days. The 3.7cm deck gun was ripped off the mounting and washed overboard, the snorkel mast bent, decking was severely bent, and much of the Alberich was ripped away. Portsmouth had considered *U-1105* and its crew lost. Soon communication was reestablished and a tug was requested to meet the *U-1105* given the mechanical state that the U-boat was in.

Murphy recalled long after the war that his command of the *U-1105* during the winter crossing of the North Atlantic was the most memorable of all his military career experiences. His efficiency report issued in March of 1946 reflected his command of *U-1105* well during the crossing:

> Lieutenant Commander Murphy is an able and experienced Naval officer. His standards of personal and military character are very high. He recently completed the difficult assignment of bringing the Ex-German U-boat *U-1105* to the US from Portsmouth England. His tactful manner and pleasing personality enabled him to enlist the full cooperation of the British Navy while fitting out the submarine and training the crew. He exhibited fine seamanship in making passage across the Atlantic in the latter part of December in the face of the season's heaviest gales without damage to his ship or injury to the crew. He is recommended for promotion when due.[74]

For the efforts of Murphy and his prize crew the US Navy's Bureau of ships determined in January 1946 that ex-German U-boats would be made available for depth charge testing by the Portsmouth Naval Yard. These U-boats must be preserved to a certain extent "to ensure that they are capable of static submergence at the time of the tests."[75]

CHAPTER 6

SALVAGE TRAINING AND DEPTH CHARGE TESTING IN THE POTOMAC RIVER

U-1105 remained at Portsmouth only for a few weeks before onward transfer to Boston. On February 1 the US Navy Bureau of Ships instructed the commander of the Portsmouth Naval Shipyard to send 0.84sq m (9sq ft) of the *U-1105*'s Alberich to the Naval Research Laboratory in Washington, and a like amount to Dr H. Bolt of the Massachusetts Institute of Technology's (MIT) Acoustic Laboratory in Cambridge, Massachusetts for testing. Also removed for study were the ship's depth gauge and other instruments having luminous backings.[76] A follow-on memorandum dated "February 1946" from the Chief of Naval Operations stated that upon arrival in Boston *U-1105* was to be preserved in preparation for the explosive tests. Also revealed in the memo is the fact that the Naval Research Labs, which was given a piece of Alberich prior to the arrival of *U-1105*, tested it with the result that "NRL investigation reveals rubber coating of British design for sound not radar protection."[77]

Figure 41. USS *Windlass* (ARS(D)-4) in the Potomac River during salvage and towing training on the *U-1105* on September 21, 1948. (Still image taken from the training film made that day)

It appears that the communication might have confused the rubber samples received from the Royal Navy for the testing at the NRL as being "British", but these samples were of German Alberich and taken from the surrendered *U-485*.

The sample that went to MIT was destined for an entirely new lab, the first of its kind in the USA. Dr Bolt proved to be a man of high intellect with an extraordinary interest in acoustics. In 1939 he one of only three American recipients of the National Research Council post-doctoral Fellowship in Physics in the same year he received his PhD from the University of California at Berkeley. His thesis, "Normal Modes of Vibration in Room Acoustics" would eventually lead him to MIT in 1940, where he directed its Underwater Sound Laboratory through 1942. Under his direction, the laboratory began to catalogue the underwater signatures of ships. This research soon drew the interest of the US Navy in the wake of the German Operation *Paukenschlag*. In 1943 Bolt was named Scientific Liaison Officer in Subsurface Warfare to the Office of Scientific Research and Development in London, where he played a significant role in coordinating the US and British response to the U-boat threat. Bolt developed the first underwater microphone. When the Royal Navy suspected that the Germans had deployed acoustic torpedoes for the first time, they reached out to the US Navy for assistance in identifying what sounds might set them off. The US Navy in turn, looked to Bolt for answers.[78]

Figure 42. USS *Salvager* (ARS(D)-3) in the Potomac River during salvage and towing training on the *U-1105* on September 21, 1948. *U-1105* is being held afloat by pontoons hooked to compressor air hoses. (Still image taken from the training film made that day)

In the spring of 1945, Dr Bolt left London and returned to MIT. In the fall of that year – just as testing of U-boats ramped up in the immediate postwar period – Bolt organized an innovative interdepartmental Acoustics Laboratory and was appointed its Director. This new department that Bolt conceived consisted of a laboratory that melded physics, electrical engineering, architecture, mechanical and aeronautical engineering, psychology, and the arts. A supervisory committee was drawn from the first three of those fields to provide policy guidance. The Acoustics Laboratory was supported primarily by funding from the US Navy's Bureau of Ships. This laboratory lasted for 12 years before government funding was terminated. At its height the laboratory employed more than 80 persons, had 12 faculty members, and made a significant dent in the literature on acoustics.[79] What testing Bolt conducted on the piece of Alberich from *U-1105* isn't known for certain, but given the nature of the lab, it was likely it involved identifying the harmonic resonance and cataloguing it for the US Navy.

A week after reaching Boston, *U-1105* was transferred to Hingham Shipyard and berthed next to *U-505*. *U-1105* was there to be retained in a state of preservation for explosive tests, along with four other U-boats, after all spare parts and equipment had been cannibalized.[80] Not until the early summer did activity begin anew.

Figure 43. Starboard side of *U-1105* while being towed on September 21, 1948. This is a good representation of what can be seen of the wreck site on the bottom of the Potomac River above the mud line. (Still image taken from the training film made that day)

Figure 44. Port side of *U-1105* while being towed on September 21, 1948. This is a good representation of what can be seen of the wreck site on the bottom of the Potomac River above the mud line. (Still image taken from the training film made that day)

Memos dated June 5 and 19 exchanged between CDRs Portsmouth and Boston noted that *U-1105* was inspected and a list of parts to be removed for use as spares or exploitation was made.[81] These spare parts were destined for use in the maintenance of the ex-*U-2513* now in US Navy service. The parts were removed and any further preservation work completed. On August 9 *U-1105* was determined ready for "disposal".[82]

A final testing plan was submitted by the Bureau of Ships that was approved on November 29 by Chief of Naval Operations that read in part:

1. The proposal contained in reference (a), that the Bureau of Ordnance conduct a series of tests on a German submarine using a new demolition outfit which would end in destruction

Figure 45. Close-up view of the starboard side of U-1105 while being towed on September 21, 1948. In this picture the Bali aerial (FuMB 29) marked (1) is mounted and still attached. It is not present on the wreck site today. Also, the unique watertight tube can be seen (2). This is discussed further in chapter 9. (Still image taken from the training film made that day)

beyond repair, and that subsequently the Bureau of Ships investigate various types of equipment and methods for salvaging submarines, are approved.

2. The ex-German submarine *U-1105*, is hereby designated for this project.

. . . 4. At the conclusion of the proposed tests, the *U-1105* shall be finally disposed of by sinking in waters of such depth as to assure a swept depth of at least fifty feet.[83]

However, delays in the testing program saw *U-1105* towed to the Washington Navy Yard instead, where it remained for nearly 18 months. On December 18 a memo from Commander, US Naval Base Boston to Commander First Naval District stated *U-1105* was to be moved from Boston to US Naval Mine Warfare Test Station, Solomons, Maryland with an arrival date there no later than January 15, 1947.[84] During 1947, *U-1105* was towed from Boston to Solomon, MD, followed by the Washington Navy Yard, where it remained for some 18 months.

During the summer of 1947 discussion first emerged about a future potential test. The Chief of the Bureau of Ordnance sent a letter to the Chief of Naval Operations on July 3 looking for a former German U-boat to conduct a "near-ship probing experiment" as part of the Underwater Electric Potential (UEP) Fields study that was concerned with the measurement of "galvanic current" in the potential distribution of water near the U-boat at the time of a detonation. A shallow depth of 24m (80ft) in a non-ocean environment was required. The response came back from Chief of Naval Operations on August 18 that no U-boats were planned to be moved any time soon but that any of the former German U-boats now in possession of the US Navy could be utilized when the time came.[85] No UEP test followed that year.

U-1105 was subsequently ordered towed from Naval Gun Factory, Washington DC to the Ordnance Test Facility, Solomons, MD about August 2, 1948.[86] The USS *Hoist* arrived at the Washington Navy Yard on August 11 after departing the Solomon the day before. It quickly hooked up *U-1105* and departed with the U-boat in tow on the port side and proceeded to Piney Point, MD. On the 12th the *Hoist* arrived at its destination in the Potomac and transferred the *U-1105* to the USS *Salvager* (ARS(D)-3) then departed back to the Solomons, its role in the *U-1105* testing complete.[87] Along with the *Salvager* was the USS *Windlass* (ARS(D)-4), both of which had arrived from Norfolk. The two vessels were reclassified after commissioning as Salvage Lifting Vessels equipped to raise submarines and conduct various other salvage operations.

U-1105 remained surfaced off Piney Point through the 16th. On either the 17th or 18th, *U-1105* was sunk in 18.6m (61ft) of water in a static drop to the bottom off Piney Point.[88] *U-1105* remained submerged through the 26th, when it was raised by the *Windlass*. Four days later the approach of a hurricane caused *U-1105* to be dropped back to the bottom off Piney Point on August 30.

On September 2 the *Salvager* raised *U-1105* off the bottom. Members of the Bureau of Ships were on hand to observe the event and take photographs. On the 21st, towing tests were conducted by employing collapsible pontoons that placed *U-1105* in a "submerged condition."[89] A black-and-white film was made of this floating process. On the 27th, "Two USN officers, and one scientist, one mechanical engineer and one consultant from the Naval Ordnance Laboratory at White Oak" reported aboard the *Windlass* for temporary duty along with five enlisted men.[90] Late in the afternoon on the 29th the first depth charge test was conducted

Figure 46. View of *U-1105*'s port side during final preparations for depth charge detonation testing on September 19, 1949. The multicolored poles allowed for accurate measurement of the blast effect on the U-boat. (Courtesy of Maryland Historical Trust)

off Buoy 16 AA in 18 fathoms off Point No Point sinking *U-1105* to the bottom for the third time.

There was no "secret" to these experiments. A *Washington Post* article dated October 12, 1948 read in part: "A former German submarine was sunk in the Chesapeake Bay September 29 in tests to determine the lethal range of depth charges, a naval spokesman disclosed today. The submarine, the *U-1105*, was sunk off Piney Point, Md. The Navy now is testing new equipment and techniques for raising a sunken submarine, using the *U-1105* as the guinea pig."[91]

U-1105 remained submerged through November while an array of salvage tests were conducted that included placing divers from the *Salvager* down on to its decks on November 5 and 7.[92] With the onset of winter, all salvage tests were ceased. The *Windlass* marked *U-1105*'s position with three yellow and one white buoy and departed for Norfolk along with the *Salvager* on the 13th.[93]

U-1105 remained on the bottom for nine months. Between July 11 and 17 divers were dropped on to the sunken U-boat to conduct an inspection. When it was sunk the prior year the inner hull was not flooded, just the ballast tanks. There was some belief that the U-boat was going to be found still airtight. When the divers opened the aft-torpedo room hatch, then the conning tower hatch, the U-boat was found to be flooded. A further inspection of the forward torpedo room confirmed this. However, both the aft-torpedo room and control room could be made airtight and the water pumped out. This could not be achieved in the

Figures 47 and 48. Two views of *U-1105*'s conning tower during final preparations for depth charge detonation testing on September 19, 1949. Of interest is that the US Navy added a valve to seal the portside snorkel intake trunking. This was likely early on during salvage testing when they dropped *U-1105* to the bottom of the Potomac River on November 13, 1948 in the hope that it would remain watertight over the winter months. Also of note in the right photo is the presence of the snorkel mast at the time of its final sinking. Unfortunately river mud has covered up the snorkel well and divers cannot access the actual snorkel mast. A probe was used by the author in a 2015 dive to confirm that the snorkel mast was still present. (Courtesy of Maryland Historical Trust)

Figure 49. The 250lb HBX-2 loaded MK2 depth charge being prepared for detonation below *U-1105*. (Courtesy of Maryland Historical Trust)

forward-torpedo room, which could only be pumped out once it was raised to the surface. During the period of July 18-31 divers lashed 10 x 40 ton and 15 x 25 ton lift pontoons and rigged air tanks across the hull of *U-1105*. Hoses were connected to the ballast tanks and all necessary valves that had been damaged during the winter months were replaced.

On August 1 the first lift operation ended in failure. The lift pontoons were filled with air first, then the ballast tanks were pumped. *U-1105* broke free from the bottom and breached the surface with such force that a wave was generated equal to the deck of the surface vessel conducting the recovery. As *U-1105* settled back down on the surface ten pontoons broke free and the U-boat descended to the bottom. Once it was considered safe, divers went back down and found that *U-1105* was listing to starboard by 10–15 degrees and that it had settled about 4.6m (15ft) into the mud.[94]

Work commenced on resetting the pontoons from August 13 through the 17. Fewer were used on the second attempt. Only 6 x 40 and 6 x 25 ton pontoons were applied, as it was determined too many caused *U-1105*'s previous rapid, yet uncontrolled ascent.[95] On August 18, *U-1105* was successful raised again. The U-boat was towed to Piney Point and sunk again on the 19th in 10.7m (35ft) of water.[96]

On September 2, *U-1105* was raised and floated for the fifth and last time. The U-boat was now readied for an "Explosive and Structural Test" by the new Underwater Explosive Research Division founded in 1946. This new US Navy department learned that during WWII underwater explosions initially produced a condensed high-pressure airwave, followed by several successive shockwaves that were attracted to underwater objects. These successive shockwaves were often the cause of significant damage to vessels and crew alike, frequently resulting in destruction and death without a direct hit being required. *U-1105* became one of their first test subjects.[97]

By the end of September 16 the preparations had been finalized. A 250lb HBX-2 loaded MK2 depth charge was suspended 9m (30ft) below the keel at the port edge, just forward of the conning tower. On September 19 at 12:00pm the detonation occurred, resulting in a split in the starboard side outer hull as *U-1105* was lifted into the air by the concussion. The U-boat then sunk in 20 seconds to its present position off Piney Point, settling upright in the soft mud of the river bottom in 27.7m (91ft) of water.[98] While in US Navy custody, *U-1105* was sunk six times and raised five, including the uncontrolled breach of the surface on August 1.

Figure 50. *U-1105* in the Potomac River facing approximately southeast. The US Navy has affixed poles with flags that will be used in measuring the force of the concussive wave produced by the proximity depth charge. (Time-elapsed image of the final detonation and sinking of *U-1105* on September 19, 1949 from the US Navy film *Close is Near Enough*)

Figure 51. The initial detonation has occurred, sending a pressure wave outward as noted by the water plumes both port and starboard of *U-1105*. (Time-elapsed image of the final detonation and sinking of *U-1105* on September 19, 1949 from the US Navy film *Close is Near Enough*)

Figure 52. The force of the initial detonation wave now raises the center section of *U-1105* out of the water. (Time-elapsed image of the final detonation and sinking of *U-1105* on September 19, 1949 from the US Navy film *Close is Near Enough*)

Figure 53. As *U-1105* starts to settle back down, the secondary pulsation wave strikes its keel, as noted by the dark clouds of explosive gas. It was at this moment that *U-1105*'s outer hull on the starboard side cracked. It has never been confirmed that *U-1105*'s pressure hull was also cracked in the detonation, though this seems very probable. U-boats proved highly resilient to depth charges during the war. For example, coroners determined that *U-853*'s crew were killed by blunt force trauma long before the pressure hull was cracked during its final engagement on May 6, 1945 off Rhode Island. (Time-elapsed image of the final detonation and sinking of *U-1105* on September 19, 1949 from the US Navy film *Close is Near Enough*)

Figure 54. The secondary and devastating pulsation wave continues its work on *U-1105*'s hull as it begins its final descent to the bottom of the Potomac River. (Time-elapsed image of the final detonation and sinking of *U-1105* on September 19, 1949 from the US Navy film *Close is Near Enough*)

Figure 55. The pulsation wave subsides and *U-1105* disappears. (Time-elapsed image of the final detonation and sinking of *U-1105* on September 19, 1949 from the US Navy film *Close is Near Enough*)

Figure 56. *U-1105*'s final moments on the surface, September 19, 1949. Two US Navy Officers record the detonation time as the initial pressure wave emerges from beneath *U-1105*. (Courtesy of Maryland Historical Trust)

CHAPTER 7

FORGOTTEN BUT NEVER LOST

U-1105 was left on the bottom to be forgotten but not lost. A wreck symbol was added off Piney Point on NOAA nautical chart #12285. On August 8, 1951 the Nelson Salvage and Construction Company of Baltimore approached the US Navy and requested permission to salvage the *U-1105*. The US Navy refused without a reason.[99] Locals recalled the testing and sinking with fading memories. Not until the accounts of its sinking were published in the 1990s after its "discovery" by sports diver Uwe Lovas on June 29, 1985 that interest in this unique U-boat was renewed. Lovas observed on his first dive that it was likely not visited by another diver since its final sinking. However, it has been reported to this author that local divers did dive the wreck before Lovas, though they may not have realized what it was they saw on the bottom.[100]

Much has been made about the fact that *U-1105* was "lost" because the report dated October 19, 1949 filed by the US Naval Ordnance

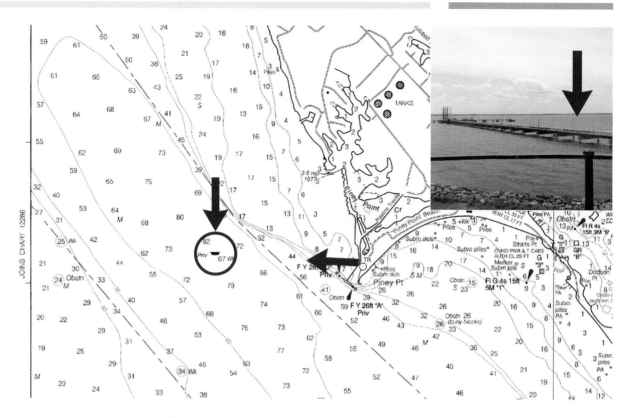

Figure 57. *U-1105*'s location was marked by a wreck symbol on NOAA charts after it was sunk. The arrow in the upper left photo depicts general location of *U-1105* dive site as related to the loading/unloading pier. The arrow in the lower image depicts the direction of view of the inset photo.

Disposal Unit and School on the U-boat's final sinking noted its position as 38°07',33N 67°32',32W, apparently transposing the longitude to read 67°W instead of the correct 76°W. While this was certainly the case, every associated document to the testing stated "Piney Point" as the location of the sinking. No one looking at the documents would have been confused that the *U-1105* was anywhere else but in the Potomac River. While there was indeed a clerical mistake made, this transposition in numbers had little to do with the *U-1105* being "lost". It certainly has made for an enduring myth.[101] *U-1105*'s final resting place was destined to be located and publicized, regardless of the circumstance.

The State of Maryland quickly took steps to make the site a Marine Preserve in the early 1990s. An initial survey of the wreck site was conducted by Michael Pohuski and Donald Shomette, their results published in a 1994 report. They inherently understood the unique characteristics of *U-1105* and wrote the following in the summary to their report:

> The *U-1105* BLACK PANTHER is one of the most unusual shipwreck sites in Maryland waters, if not the entire Atlantic seaboard. The boat is archaeologically, historically and technologically significant. As one

of only ten Alberich vessels built for the German Navy during World War II to neutralize sonar detection, she represents one of the earliest successful efforts of modern stealth technology to be fielded in wartime. Of the original 660 Type VIIC Class boats built, there is only one other preserved example surviving, in Kiel, Germany.

Architecturally the vessel is important because it differs in several ways from the known standards of the Type VII-C submarine design from which it sprang. The most visible components that reinforce this assertion include: the two large portals which provided access to the lower Wintergarten gun pedestal; the semicircular flange or platform extension on the starboard aft side of the upper gun deck; the design of the starboard con ladder; the GHG Balkon passive sonar system buried below the vessels bow and of course, the rubber-cladding of the entire ship.

Historically, the vessel's military role in World War II and its subsequent utilization in US Naval weapons testing programs in the field of antisubmarine warfare is unique. The impact of "antistealth" technology developed as a consequence of US Navy tests conducted on the *U-1105* have undoubtedly played an important role in the promulgatio n of naval technological inquiry, development, and possibly even policy during the Cold War which only further research can reveal.

The maritime archeologists who surveyed the wreck clearly understood the importance of *U-1105* historically, if they did not divine all the nuances of its past and engineering uniqueness. They did not remark on the value of the snorkel mast buried beneath the silt, along with its hydraulic lift mechanism. They noted the port side snorkel trunking that

Figure 58. Right. A top-down view sidescan sonar image of *U-1105*'s final resting place on the bottom of the Potomac River. (Courtesy of Lighthouse Archaeological Maritime Program)

Figure 59. Far right. Sidescan sonar image of the starboard side of *U-1105*. This image shows just how little of the U-boat protrudes from the river mud. (Courtesy of Institute of Maritime History)

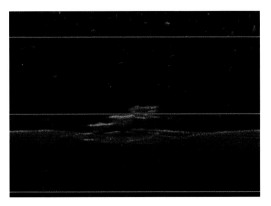

ran along the conning tower just below the level of the upper Wintergarten's base, but they did not understand how the system worked. They documented, but did not identify the starboard side exhaust trunking and emergency shut-off valve. As important as the Alberich was, its deployment by the BdU after the introduction of the snorkel was not understood within the historical context of Dönitz's "Total Undersea War".

The only point of debate that remains about *U-1105* is why the US Navy risked a crew to sail it across the North Atlantic in the middle of winter. Perhaps some document will be revealed at a future date that confirms the rationale for this decision. It seems that if the Western Allies simply wanted to keep it out of Soviet hands then they could have just as easily sunk it in British coastal waters during Operation *Deadlight*. The near sacrifice of a US Navy prize crew combined with the unique qualities of *U-1105* easily earns it the distinction of being a "National Treasure" that deserves to be raised, restored and put on display as in the case of *U-505* (IXC) in Chicago's Museum of Science and Industry, and not to be silted over at the bottom of the Potomac River.

In museums across the world – from Washington, DC, to London and Moscow – sit numerous examples of German World War II technological innovations such as jet fighters and rockets that propelled the next generation of engineering and development in their fields. *U-1105* should be raised and restored as an example of the transition from the prewar submersible to the modern submarine. It would be a testament to the sacrifices Allied sailors made against a dogged enemy they fought to the bitter end.

Figure 60. Below left. Updated rendering of *U-1105*'s starboard side as it appears in the Potomac River mud circa 2017.

Figure 61. Below right. Updated rendering of *U-1105*'s port side as it appears in the Potomac River mud circa 2017

CHAPTER 8

DIVING THE *U-1105* TODAY

Diving the *U-1105* is not difficult, but it is also not for the inexperienced diver. It is dark on the dive site and currents can be experienced. Any disturbance of the silt can cause a total blackout, even with the most powerful lights. Divers must be comfortable functioning independently in the dark and be able to self-deploy a surface marker buoy (SMB).

There are two buoys on the sub. The large (1.8m/6ft tall) white buoy is located near the sub. A small (diver ball) buoy is tied directly to the conning tower. Descents and ascents typically occur on the line of the smaller buoy.

The depth to the top of the conning tower is about 20m (65ft). The depth to the silt is about 26m (86ft).

The sub is located at 38°08'.10N 76°33'.10W. This is 1 mile west of Piney Point, Maryland. The Potomac River is quite impressive at the point where the sub rests. The river is about 2½ miles across and about 27.5m (90ft) to the bedrock.

The visibility ranges from less than 0.9m (3ft) to as much as 3.7m (12ft) depending on the time of year, water temperature, and rain in the Chesapeake Bay watershed. Due to the particulate matter in the water, the sunlight rarely penetrates to the conning tower. The sediments on and around the submarine consist of a black-brown, very fine river bottom silt that is quite fluid. If these sediments are disturbed, a black cloud forms instantly around the diver. Disturbing the silt on or near the wreck will effectively make the *U-1105* temporarily a "zero-vis" dive. However, currents on the site will usually remove the cloud in a few moments. If there is not enough current to quickly disperse the silt, ascend 1.5m (5ft) or so to get above the sediment cloud.

The water is tidal and brackish. The river water mixes with the Chesapeake Bay salt water and ranges between 6 and 18 parts per thousand (ppt) salt. Drinking water has less than 0.5 ppt salt, while the oceans have about 33 ppt salt.

Average surface temperatures range from 3.9ºC (39ºF) in February to 27ºC (80ºF) in July. Bottom temperatures can be 5ºF or so less than surface temperatures. Strong currents are possible. The best time to dive is at or near the high tide. However, there are usually counterflowing top and bottom currents of different intensities.

The Maryland Historical Trust diving guidelines recommend that all dives be no-decompression dives. It is recommended that each diver carries a reel and lift bag/SMB. A big light and a back-up light are also a requirement. When you are on the U-boat, and your light goes out, there is a complete absence of light. Fishing line can be present across the wreck so multiple cutting tools are recommended.

The dive boats that operate here are not typically the ones utilized for ocean wreck diving. These boats are small and do not have a sophisticated entry/exit system, or even a platform. Divers will typically roll off the boat at the start of a dive and be required to doff their gear in the water and climb back into the dive boat after the dive.

Commercial and pleasure boat traffic on the river is a concern. Therefore, it is recommended that dive boats use both the official ALFA flag (blue and white) and the unofficial sport diver flag (red with white diagonal). Both flags are required under Maryland law.

Look but don't touch. *U-1105* is part of the US National System of Marine Protected Areas. Many features and artifacts remain on the

Figures 62, 63 and 64. The orange dive ball typically marks the conning tower of *U-1105*. This acts as both the descent and ascent line to *U-1105*. The *U-1105* "Black Panther Maritime Preserve" marker buoy is anchored by a chain to a clump that rests upon the river's bedrock off the port bow of the U-boat. It is typically used to anchor the dive boat. Each winter season it is removed and replaced again in the spring. The author is pictured rolling off the side of the dive boat *Audrey B* in 2017. Diver Fred Engle can be seen in the upper right making the swim from the boat to the orange dive ball. (Images courtesy of Kameron Hamilton and Fred Engle).

U-1105. Some features are fragile and all objects are US Navy property; nothing may be removed from the site. Wreck penetration and artifact recovery are strictly prohibited: offenders will be prosecuted under US Federal Law as per the 2004 Sunken Military Craft Act (SMCA).

CHAPTER 9
ARCHEOLOGY OF A LATE-WAR U-BOAT

The U-boat rests on its keel with almost no list. Only the conning tower, upper and lower Wintergarten, and a small part of the hull are out of the silt. The only portion that can be seen above the silt is the conning tower itself, with another 5–7.5cm (2–3in.) of the main deck forward of the conning tower and 10–15cm (4–6in.) aft of the Wintergarten. Despite these limitations, nowhere else on the North American coast can a diver see a snorkel- and Alberich-equipped U-boat.

The following images provide a guide to what remains of *U-1105*. The preceding chapters have provided a concise overview of the *U-1105*'s history and technological innovations. The pictures below reveal what is left on the bottom of the Potomac River and in the Piney Point Light House and Museum. Only through years of extensive archival research can these features now be understood within their appropriate historical context.

Of all the prominent features that remain intact and accessible for view to divers is the snorkel trunking as discussed in Chapter 1 and labeled in the following images. This feature, more so than Alberich, gave U-boats the ability to continue to be effective as a weapon system against the Allies in the last year of war. It was the snorkel system that evolved U-boats from prewar submersibles to late-war submarines and affected all later submarines designs the world over. When German U-boat crews remained underwater for a week, then a month, then two months at a time they were setting records and testing the psychological and physiological limits of human endurance soon duplicated by future submarine crews the world over.

The snorkel proved the most sought after and copied U-boat technological innovation in the postwar period. While the US Navy ignored Alberich for almost 40 years, it adopted the snorkel immediately and designed the Greater Underwater Propulsion Power Program (GUPPY) modernization program around this device. The US Navy's 1961 edition of its submarine technical training manual known as *NAVPERS 16160-B The Submarine*, issued to all crew members of the new GUPPY-modified submarines, offered unusually high praise to their former German enemy less than 20 years after the end of WWII with the following commentary on the snorkel. The Introduction to Chapter 15's "The Snorkel System" reads: "The theory of the snorkel had been known for several years; but, it was not until 1943 that the German Navy converted such theory into practical operation. . . . the German Navy perfected snorkel designs and incorporated the device in their submarines. This move increased the efficiency and success of German underseas craft immeasurably."[102] The GUPPY program also included additional German wartime innovations such as increased battery capacity and a streamlined hull. However, when funding began to become scare for the program, the US Navy chose to continue retrofitting its fleet with the snorkel at the expense of other modifications.[103]

This study provides for the first time the compete context of the exposed port and starboard side trunking. Divers now have a clear understanding of what they are seeing in the blackness of the Potomac River bottom and its importance to postwar submarine development.

The Alberich is also a key feature. It can be seen throughout the conning tower and lower along the hull. A close examination will reveal

Figure 63 The enlarged front view of the conning tower represents what is generally viewable above the river mud. Key features are marked.

1. Starboard side snorkel exhaust trunking
2. Emergency exhaust snorkel shutoff vale
3. Magnetic compass
4. Access hatch to tower casing
5. Compressed air whistle/horn
6. Port side snorkel (Type 1 flange) air intake
7. Starboard side navigation light
8. Spray deflector
9. Port side snorkel clamp (when raised)
10. Wind deflector
11. Attack periscope housing
12. DF aerial housing
13. UZO binocular mount
14. Aerial periscope housing
15. FuMO 61 Hohentwiel radar housing
16. Forward radio antenna outlet

differences in its pattern according to its location. It is a testament to German wartime engineering and study of the application of sound in an underwater environment that exceeded that of every other nation at that time. It was measurably effective, as the account by the 21st Escort Group's first and only interaction with *U-1105* shows they never detected the U-boat, even though *U-1105*'s evasive maneuver likely brought it directly under two of the group's destroyers while they were using active ASDIC.

As previously noted, after the war Alberich was adapted by the Soviets for their submarines as early as the 1950s. However, the US Navy was never that interested in this specific German technology. They studied it, then all but forgot about it until the late 1970s. Acoustic tiles did not appear on a US Navy

Figure 64 The enlarged front view of the conning tower represents what is generally viewable above the river mud. Key features are marked. Note that all guns and railings were removed.

1. Lower Wintergarten
2. Starboardside snorkel exhaust trunking and emergency shut-off valve
3. 3.7cm watertight ammo container
4. 2cm watertight ammo container
5. 2cm gun platform
6. Upper Wintergarten
7. Starboardside diesel air intake (non snorkel)
8. Starboard side navigation light
9. UZO binocular mount
10. Wood panelling
11. Aerial periscope housing
12. DF aerial housing
13. Sky periscope housing
14. FuMO 61 Hohentwiel radar housing

submarine until 1980, almost 30 years after their Soviet counterparts had adopted the technology.

Other features of interest also exist as referenced in the following photographs. Note that all underwater images have been labeled and marked on two drawings rendered by the author of the wreck site as it looks today. These images will allow any diver who studies them to be able to quickly identify the features of *U-1105* exposed on the wreck, and hopefully make their experience in the black waters of the lower Potomac River that much more memorable. Unfortunately the wreck's condition has deteriorated over the decades and the river mud has swept over much of the decking that was once exposed back in the 1980s and 1990s.

LOCATION OF UNDERWATER PHOTOS ON THE WRECK

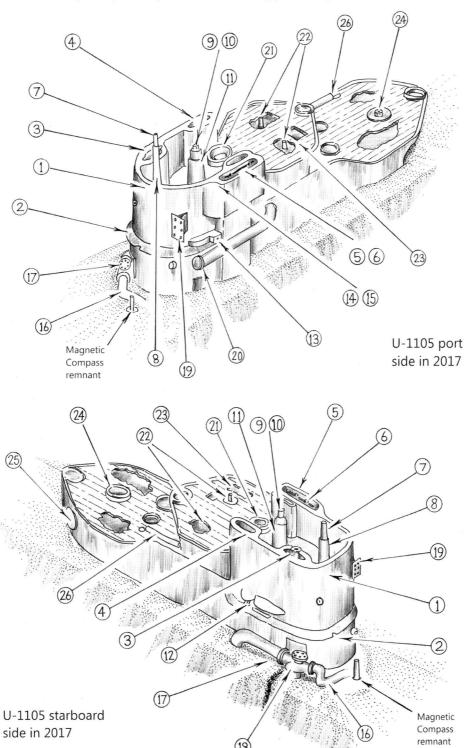

U-1105 port side in 2017

U-1105 starboard side in 2017

(1) *U-1105*'s conning tower as viewed from the port side. The wind deflector is almost gone, with a small piece remaining on the right side. Rising up from the center left is the sky periscope. Bottom center can be seen the mounting bracket for the snorkel clamp. (Courtesy of Paul D. Lenharr II)

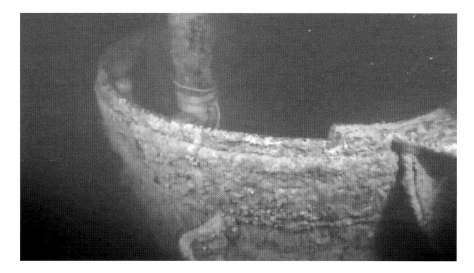

(2) Halfway down the front of the conning tower is the wave guard that remains surprisingly intact. (Courtesy of Paul D. Lenharr II)

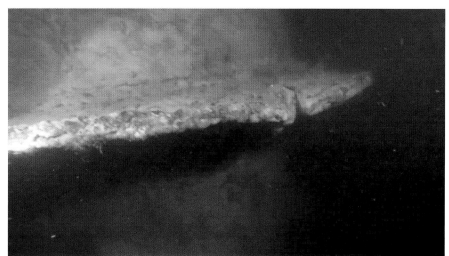

(3) On the starboard side of the conning tower is the raised universal mount that could accommodate the interchangeable AM antenna and round dipole. The round dipole is still present in the recess to the right of the mount. (Courtesy of Paul D. Lenharr II)

(4) Just behind the raised mount are the starboard side diesel engine air intake (left) and the 2cm watertight ammunition container (right). (Courtesy of Paul D. Lenharr II

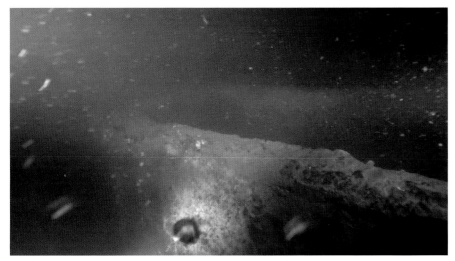

(5) On the port side of the conning tower is the FuMO 61 Hohentwiel radar recess. (Courtesy of Tom Edwards)

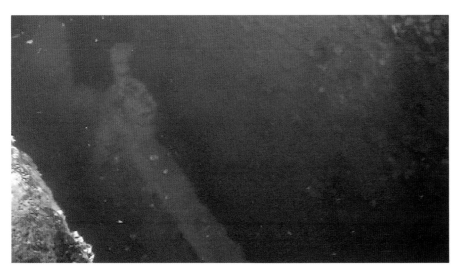

(6) The FuMO 61 Hohentwiel radar is still present within the recess. The top of the radar detector can be seen in the image. (Courtesy of Tom Edwards)

(7) The top of the sky periscope mount. Fishing line has been caught around this feature as it sits higher off the bottom than any other part of U-1105. (Courtesy of Tom Edwards)

(8) Base of the sky scope. This feature is often used alternately with the attack periscope housing to anchor the orange dive ball. Rope can be seen tied along its base. On the left side of the image you can see the starboard interior of the conning tower. The actual wood paneling is still present underneath the layer of mud and can still easily be seen with a brief swipe of a gloved hand. (Courtesy of Paul D. Lenharr II)

(9) Top of attack periscope mount. Its burnished steel still reflects the illumination from a dive light. In this image, rope has been tied to anchor the orange dive ball on the surface. (Courtesy of Tom Edwards)

(10) Top of attack periscope mount. This image offers a clear perspective of the burnished steel. (Courtesy of Tom Edwards)

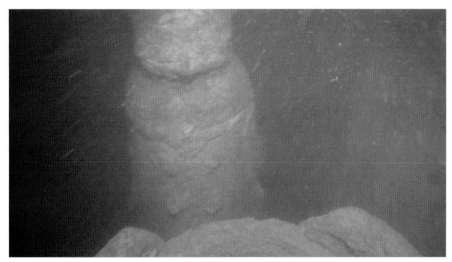

(11) Bottom of attack periscope mount. Mud and concretion has begun to build up, though it still retains its shape. (Courtesy of Tom Edwards)

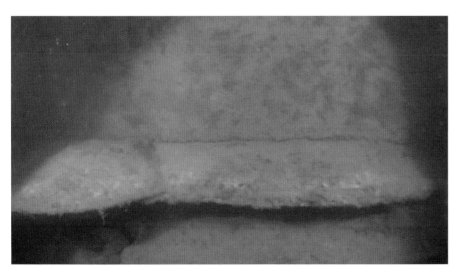

(12) Starboard side light housing (left). (Courtesy of Paul D. Lenharr II)

(13) Port side light housing. (Courtesy of Paul D. Lenharr II)

(14) Alberich along the top of the port side of the conning tower. This image clearly details the two-ply design of the antisonar coating. (Courtesy of Fred Engle)

(15) Alberich consisted of a pattern of 2mm and 5mm holes arranged in various patterns that were specific to each section of the U-boat, depending on the hull thickness and whether the backing was water or air. In this case, this section of Alberich does not have any discernable pattern. The hole pictured here appears to have been formed by a bolt that affixed it to the conning tower. (Courtesy of Fred Engle)

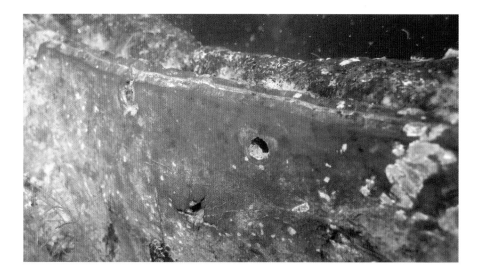

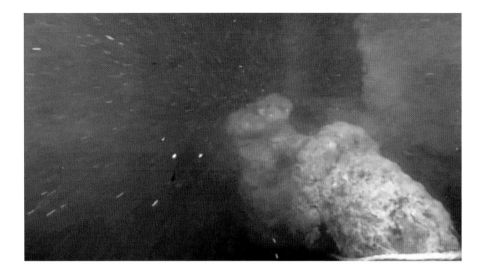

(16) Starboard side exhaust shut-off valve looking aft. Note the round flat top that provides access to the valve from outside of the U-boat. The original maritime archeological survey conducted by the Maryland Historical Trust never identified this raised feature that ran above the deck and connected the diesel engine exhaust outlet to the snorkel exhaust outlet. The exhaust trunking ran above the starboard side deck on all Type VIIC and VIIC/41s. It contained a shut-off valve that resided in the control room that allowed the crew to shut off the exhaust and stop the backflow of either water or toxic gas into the engine room when the snorkel head was inadvertently submerged during operation. It was a critical and unique component to all snorkel-equipped Type VIIC U-boats. (Courtesy of Paul D. Lenharr II)

(17) This composite illustrates the starboard side exhaust trunking, looking down/forward as it wraps around the Conning Tower. The top of the shut-off valve is detailed and shown connected to the aft and fore exhaust piping. Ten bolts normally affixed the cover plate to the exhaust shut-off valve. The cover plate measures 8.5 inches across and was removable when maintenance of the valve mechanism was required. Note that the bolt at the 1:00 position is missing having likely been blown off during the final depth charge test. There is a near symmetrical 3/4 inch hole in the middle of the shut-off valve cover plate that was not part of the original installation. It may have been drilled during salvage tests while under the care of the US Navy. The forward portion of the exhaust pipe is now cracked as it bends downward before turning to port where it connects to the snorkel mast's boot heel. No other diveable Type VIIC U-boat in North America has snorkel trunking. This is one of the few diveable U-boats anywhere in the world with intact snorkel trunking that can be viewed with ease. (Courtesy of Fred Engle.)

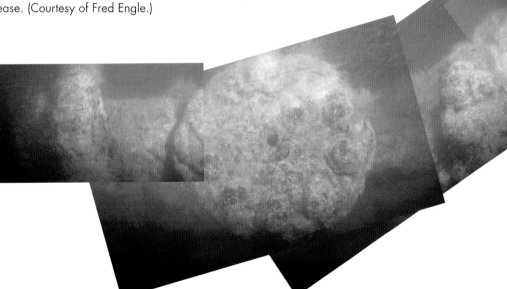

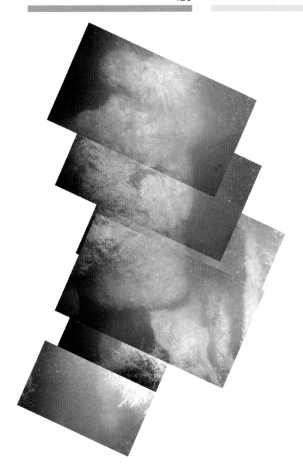

LEFT (18) This composite (looking aft) shows how the exhaust shut-off valve seats on the outer hull and its subsequent connector that runs down into the Control Room through the pressure hull where the actual turn wheel to operate the valve is located. The Conning Tower is on the right.
(Courtesy of Fred Engle)

(19) This is the base of the port side snorkel clamp. The clamp itself is missing. When the snorkel was raised it would rest within the brackets. A mechanism in the attack room of the conning tower was operated that caused a pin to lock into a static latch on the snorkel mast and secure it in place during operation. (Courtesy of Paul D. Lenharr II)

(20) Portside snorkel air intake looking aft. Midway below the top of the conning tower and about 1m (1.1yd) below the snorkel clamp is what remains of the snorkel air-intake trunking. U-1105 was equipped with the first generation Type I flange snorkel system that required both the snorkel mast and air-intake snorkel trunking to form a watertight seal between two separate flanges. This proved problematic and German engineers soon designed the Type II non-flange intake system. The actual flange valve is no longer present on U-1105. (Courtesy of Tom Edwards)

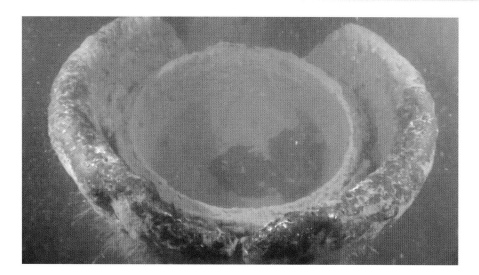

(21) 2cm watertight ammunition canister located aft of the conning tower on the upper Wintergarten (gun deck). (Courtesy of Paul D. Lenharr II)

(22) The mount for one of the 2cm Flak 38 cannon located on the upper Wintergarten. Snorkel-equipped U-boats rarely, if ever, used their anti-aircraft deck guns because they spent their entire patrol submerged. Most guns became inoperable due to rust and other mechanical failures after being submerged for a few weeks. (Image courtesy of Paul D. Lenharr II)

(23) Original wood decking is still present on the upper Wintergarten. (Courtesy of Paul D. Lenharr II)

(24) The 3.7cm automatic M42 gun mount on the lower Wintergarten. (Courtesy of Paul D. Lenharr II)

(25) One of the two storage portals along the base of the lower Wintergarten. These portals held the *Rettungsbehälter* (rescue container) that housed one five-man life raft. *U-1105* was equipped with four such pods on its forward deck. There is no evidence that *U-1105* was equipped with two additional rescue containers below the Wintergarten during its only wartime patrol. The mud line now runs up to these openings and covers the entire aft deck of *U-1105*. (Courtesy of Paul D. Lenharr II)

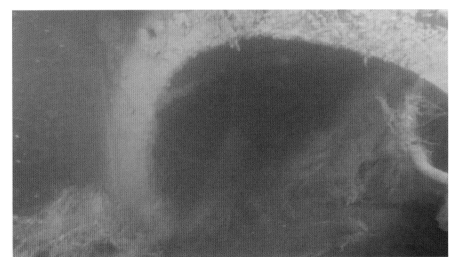

Portside light fixture and bulb from *U-1105* on display at the Piney Point Lighthouse Museum. (Author's collection)

Far Right *U-1105*'s attack periscope on display at the Piney Point Lighthouse Museum. (Author's collection)

(26) A unique feature of *U-1105* is a watertight storage tube that runs along the starboard side of the upper Wintergarten. The two upper images clearly depict its presence on *U-1105* and that it has a hinged cap to allow easy access to its contents. The below image shows it still intact with its cap still in place. The cap has not been evaluated to determine if any markings exist. (Bottom image courtesy of Paul D. Lenharr II). This tube began to appear on a number of late war U-Boats in 1944 and probably housed its flag masts.

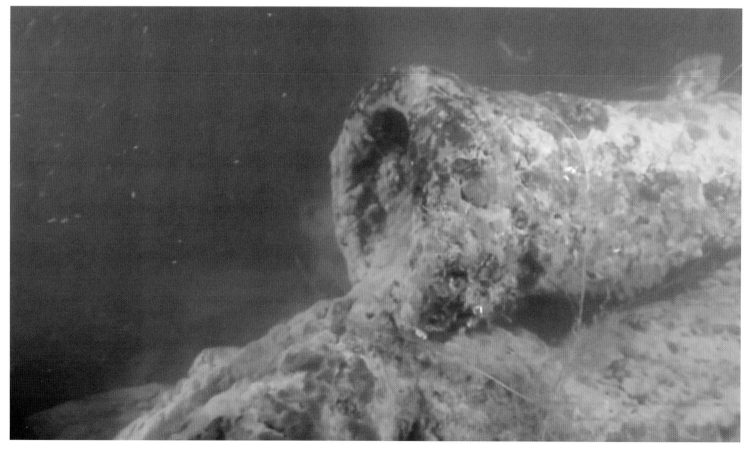

Exterior portside light housing from a *U-1105* fixture on display at the Piney Point Lighthouse Museum. (Author's collection)

Name plates removed from the conning tower of *U-1105* on display at the Piney Point Lighthouse Museum. The base of the attack periscope is visible above. (Author's collection)

Janet Murphy, daughter of US Navy Capt "Hugh" Murphy, standing next to a hatch from *U-1105* that her father likely closed during the rough crossing of the Atlantic during the winter of 1945–46. (Author's collection)

APPENDICES

APPENDIX A
Technical specifications

U-boat type:	VIIC
Displacement (tons):	769 surfaced 871 submerged 1,070 fully loaded
Length (m):	67.5 over all 50.50 pressure hull
Beam (m):	6.20 over all 4.70 pressure hull
Draught (m):	4.74
Height (m):	9.60
Power (hp):	3,200 submerged
	750 submerged
Speed (knots):	17.7 surfaced 7.6 submerged 6–8 during snorkeling
Range (miles/knots):	8,500/10 surfaced 80/4 submerged
Torpedoes:	14 4/1 (bow/stern tubes)
Crew:	44–52
Max depth (m):	c 220
Pressure hull (mm):	18.5

Equipped with:
- Turmumbau IV Wintergarten with 2 x twin 2cm Flak 38 cannons on the upper deck and a 3.7cm automatic M42 on the lower deck.
- Type I Flange Snorkel mast, with ball float and no antiradar matt on the snorkel head.
- Standard late-war Alberich configuration for a VIIC tuned to 40m (130ft) depth or less.
- GHG Balkon passive sonar array mounted under the bow, just forward of the keel.
  FuMO 61 Hohentwiel radar
  Wanze 2 radar detector/Bali aerial (FuMB 29)

Radio equipment:

20 Watt KW transmitter	3,750–15,000, (80–20m)
150 Watt LW transmitter	300–600kHz (1,000–500m)
5R/6 Kr. FKW receiver	1,500–25,000kHz (200–12m)
Direction finder:	75–1200kHz and 15–33.4kHz (4,000–250m and 20,000–9,000m)
Radio receiver:	150–441kHz and 517–21,900kHz (2,000–680m and 580–13.7m)
Late-war Kurier flash transmission system. Only the Gerber transmitter would be installed.	

Sonar equipment:

40 Watt KW transmitter	5,000–16,670kHz (60–18m)
4R/2 Allw.Empfg.	15,000–20,000kHz
GHG Balkon	

1 W.UK Tornistergerät Kriegsmarine Radio 14.67–46.15MHz (2–6.5m)
Depth sounder

APPENDIX B
The crew
Total: 52

Commander	Oblt.z.See Hans-Joachim Schwarz
1st Watch Officer	Lt.z.See Herbert Schmitt
2nd Watch Officer	Lt.z.See Heinz Sonnenrein
Chief Engineer	Ol(Ing)d.R. Kurt Larsen
Engineering	L(Ing) Armin-Norbet Schmidt-May Ob. Masch. Stolper

Crew
Ob Strm Fröhlich
Ob Masch Ekelmann
San Ob Mt Kuhne
Ob Btsmt Fidler
Btsmt Erdmann
Btsmt Trumpf
Mech Mt Danzer
Fk Mt Neisecke
Fk Mt Pascheberg
Masch Ob Mt Savelsberg
Masch Mt Hamann
Masch Mt Burchert
Ob Masch Mt Karte
Masch Mt Kreutz
Masch Mt Molitor
Mtr Ob Gefr Stank
Mtr Ob Gefr Rernhardt
Mtr Ob Gefr Wermter
Mtr Ob Gefr Wendler
Mtr Ob Gefr Weiland
Mtr Gefr Schneider
Mtr Ob Gefr Ciesinski
Mtr Gefr Steck
Mtr Ob Gefr Cherr
Mtr Gefr Cassel
Mtr Gefr Wegner Kart
Mtr Gefr Wegner Günther
Mtr Gefr Spönemonn
Fk Ob Gefr Lux
Fk Gefr Mehwald
Fk Ob Gefr Schmitt
Mech Gefr(A) Günther
Meeh Gefr(T) Franke
Mech Ob Gefr(T) Dreste
Masch Ob Gefr Steinhäuser
Masch Gefr Piecuch
Masch Gefr Niehus
Masch Ob Gefr Spenke
Masch Gefr Gößt
Masch Gefr Gebhardt
Masch Ob Gefr Plösser
Masch Gefr Krüger
Masch Gefr Patschkowski
Masch Gefr Heißmann
Masch Gefr Zwacke
Masch Gefr Henkle

Figure 98. Undated wartime image of *U-1105*'s crew. Klt Hans-Joachim Schwarz is pictured in the white cap just left of center. The watertight canister is present on the upper Wintergarten. The starboard side exhaust trunking is visible to the right. (Image courtesy of the U-boot Archiv)

APPENDIX C

Chronological history of *U-1105*

1943–44 Construction

June 6, 1943	Keel laid at Nordseewerke, Emden.
April 20, 1944	Launched in Emden. June 3, 1944 Commissioned in Emden. Alberich was likely applied in early 1944 but the historical record is not clear on this point. It received its snorkel retrofit in May 1944. This was one of the few U-boats to receive its snorkel before it was commissioned into the fleet because it was designated during construction to be coated with Alberich.

1944 Testing and training
U-Boot Acceptance Commission (UAK)

June 8	Arrived at Kiel.
June 13	First snorkel trials.
June 27	Depart for Swinemünde.

Training

June 28–30	Air defense training at Air Defense School VII.
July 2	Arrived at Danzig.
July 4–27	Deep-dive and long-transit training. Assigned to 19th U-boat Flotilla
July 28	Arrived at Pillau for dry tactical escort training, aircraft ID and general maneuver exercises.
August 4–16	Conducted five training exercises. Assigned to 8th U-boat Flotilla August
17–September 5	Overhaul of U-boat at Danzig. Assigned to 20th U-boat Flotilla
September 15–23	Pre-tactical training.
October 2–21	Dry-dock(?) in Gotenhafen. Assigned to 26th U-boat Flotilla
October 23–27	Torpedo testing.
October 27–November 6	Torpedo firing.
November 20–29	Tactical exercises in Gulf of Danzig.
December 4	Departed for Wilhelmshaven.
December 15	Arrived at Kiel. 1945
January–March	Final overhaul and equipping at Wilhelmshaven.
March 19	Departed for Kiel.
March 23–April 1	Conducted hydrophone tests and picked up all final supplies, fuel and ammunition.
April 4	Departed Kiel for passage to Norway.
April 6	Arrived at Horten.
April 6–8	Final snorkel and deep-dive testing. First and only combat patrol
April 12	Departed Horten for the Irish coast.
April 23	50 nautical miles west of North Uist along the continental shelf heading south.
April 27	Two torpedoes fired at the HMS *Redmill* then bottomed. One T5 struck the *Redmill* while the second, a LUT, likely missed.
April 27–28	Bottomed for 31 hours in 172m (564ft) of water – an extreme depth!
April 28	Surfaced and resumed patrol of area off Black Rock.
May 4	Received communication from Adm Dönitz to cease hostilities.
May 10	Arrived Loch Eriboll to surrender. British take command and transfer it to Lishally, Northern Ireland the next day. Post-war testing by Royal Navy Assigned to 3rd Submarine Flotilla
June 3	Commissioned as Royal Navy Submarine N.16. Pennant Number painted on side of the U-boat. Prepared for trials with *U-1171* "White Puma" that received Royal Navy Submarine Pennant N.19.
June 29	Arrived at Holy Loch, Scotland.
July	No testing.
August 6	Arrive Coll Sanctuary for diving and submerged trials.

August 7	Aircraft search test likely to use airborne radar against the Alberich when the U-boat was surfaced.	
August 8–11	Testing against the snorkel. Returned back to Lisahally.	
August 15–16	Likely radar signature return testing of antiradar covered and non-covered snorkel heads.	
August 21–24	Entered dry-dock at Holy Loch to repair worn Alberich.	
August 25–26	Deep dive and snorkel tests at Inchmarnock (north) and Rothesay.	
August 27–31	Side-by-side testing with *U-1171* at Loch Goil and Lisahally.	
September 4	Snorkel testing at Loch Goil.	
September 9	Cleaning of U-boat in preparation for the Tripartite Naval Commission (TNC) inspection scheduled for the 11th.	
September 10	Snorkel trials at Loch Goil.	
September 11	Snorkel trials at Holy Loch.	
September 16–17	Snorkel testing at Fairley and Rornesey, likely thermal radiation detection.	
September 18	Noise trials at Holy Loch.	
September 19	Noise trials at Fairlie.	
September 20	Noise trials at Roseneath.	
September 25	Noise trials at Fairlie.	
October 1–19	Extensive comparison testing, likely sonar and radar, with *U-1171* at Holyhead.	
October 22–mid-Nov	At Fort Blockhouse, Gosport for a complete overhaul of diesel engines.	
November 31	Final trial with *U-1171*. Transfer to the USA	
December 6–7	US Navy prize crew inspects *U-1105*.	
December 8–18	Preparations for US Navy acquisition and transatlantic crossing.	
December 19	*U-1105* departs Portsmouth, England for Portsmouth, New Hampshire. While at sea for 14 days, *U-1105* experienced four days of storms that caused heavy damage. The U-boat was under orders not to submerge, so to negotiate the heavy seas the hatches were closed and the snorkel was employed in order to bring in the air required to run the diesels. An onboard fire broke out and was extinguished. A ten-day radio outage occurred and *U-1105* was considered lost by Portsmouth, NH authorities before it arrived. This was the last German U-boat of World War II to cross the North Atlantic under its own power.	

[PIB] 1946

January 2	Arrived at Portsmouth, New Hampshire.
mid-January	Arrived at Boston and allocated for explosive testing. It was placed in a state of preservation and remained berthed next to *U-505* at Hingham Shipyard for 12 months.
February 1	Navy Bureau of Ships orders two 2.75m (9ft) sections of Alberich removed from the U-boat's hull for testing. One is sent to the Naval Research Laboratory and the other to Massachusetts Institute of Technology where its underwater harmonic resonance was likely measured.

1947
US Navy testing

January 15	Towed from Boston to US Naval Mine Warfare Test Station, Solomons MD then to the Naval Gun Factory, Washington, DC, where it remained for 18 months.

1948

August 11	Towed to Piney Point, MD.

First sinking

August 17 or 18	Sunk in 18.6m (61ft) of water in a static drop, where it remained submerged.
August 26	Raised by the *Windlass*.

Second sinking

August 30	Sunk in a static drop to avoid damage from an approaching hurricane.

September 2	Raised by the *Salvager*.	**Fifth sinking**	
September 21	Towing tests were conducted and filmed.	August 19	Towed to Piney Point and sunk in 10.7m (35ft) of water.
Third sinking			
September 29	First depth charge test conducted off Buoy 16 AA off Point No Point in 33m (108ft) of water and remained submerged through November. *Washington Post* article was published about the test on October 12.	September 2	*U-1105* was raised off the bottom and made ready for an "Explosive and Structural Test" by the new Underwater Explosive Research Division that had been founded in 1946.
November 5–7	Salvage tests conducted that include placing US Navy divers on to *U-1105*'s decks	**Sixth sinking**	
November 13	*U-1105*'s location was marked with three yellow and one white buoy and left on the bottom for nine months.	September 19	Sunk in 27.7m (91ft) of water off Piney Point, MD after a 113kg (250lb) HBX-2 loaded MK2 depth charge was detonated while suspended 9m (30ft) below the keel at the port edge, just forward of the conning tower. Its location was openly marked on NOAA nautical chart #12285 as a wreck without its identification as a U-boat.
1949			
July 11–17	US Navy divers conducted an inspection of *U-1105* on the bottom. The U-boat was sunk in an airtight condition the prior year, but US Navy divers found it flooded. Plans were made for a partial pumping of *U-1105* in preparation for it being raised.		
		1985	
		June 29	Uwe Lovas, a local Virginia diver from Fredericksburg locates *U-1105* and keeps the discovery private.
July 18–31	10 x 40 ton and 15 x 25 ton lift pontoons were rigged with air tanks across the hull.		
		1991	
Fourth sinking		November 16	Lovas introduces an author of U-boat wreck sites, Professor Henry Keatts, to the *U-1105* dive site.
August 1	The air requirements to raise *U-1105* were not calculated correctly and when the lift pontoons were filled the U-boat burst off the bottom and breached the surface with such force that it caused a wave equal to the height of the lifting vessel's main deck. After its brief surfacing *U-1105* settled back down to the bottom where it buried itself 4.6m (15ft) into the mud at a 10–15-degree list to starboard.		
		1992	
		March	A magazine article is published by Keatts about *U-1105*, publically revealing its location.
August 13–17	US Navy divers reconfigured the lifting pontoons on *U-1105*. Only 6 x 40 ton and 6 x 25 ton pontoons were applied.	**1995**	
August 18	*U-1105* was raised in a successful lifting operation	May 8	*U-1105* becomes Maryland's first historic shipwreck preserve.

ENDNOTES

The documents cited below are derived primarily from the following institutions:

USA
Maryland Historical Trust, Crownsville, Maryland
National Archives and Records Administration, College Park, Maryland

Great Britain
The National Archives, Kew
Royal Navy Submarine Museum, Gosport

Germany
Bundesarchiv-Militärarchiv, Freiburg
U-boot Archiv, Altenbrucher
Zentrales Institut des Sanitätsdienstes der Bundeswehr Kiel-Kronshagen und Schifffahrtmedizinisches Institut der Marine, Kiel

1. National Archives and Records Administration U.S Naval Technical Mission Europe, Technical Report 517-45, "The German Schnorchel", October 27, 1945, hereafter cited as NARA NTME/TR 517-45.
2. National Archives and Records Administration, Record Group 38, German Archives TAMBACH, Box T94 "Schnorchel".
3. Malorny, and Dr. Hellmut Uffenorde "Otological Experience with Snorkel-Equipped U-Boats" (?).
4. Malorny.
5. See Lecturer Dr med habill Gunther Malorny, "Carbon Monoxide on Submarines", Translated by US Fleet, US Naval Forces, Germany, Technical Section (Medical), SchiffMedInstM. Dok. Nr. 10975, Zentrales Institut des Sanitätsdienstes der Bundeswehr Kiel-Kronshagen und Schifffahrtmedizinisches Institut der Marine, Kiel, 1994.
6. Vorläufige Beschreibung und Betriebsvorschrift der U-Boot Schnorchel Anlage, Tiel 3, U-Boot Archiv, Cuxhaven, Germany.
7. Dr. Guenther Malorny, "Carbon Monoxide on U-Boats" (1994).
8. PRO/ADM 1-17549.
9. The three best sources of Type VIIC German U-boat construction that include detailed diagrams remain Eberhard Rössler's The U-Boat: The evolution and technical history of German submarines; David Westwood's The Type VII U-Boat; and Fritz Köhl and Axel Niestlé, Vom Original zum Modell: Uboottyp VII C: Eine Bild- und Plandokumentation. The line diagrams in all three books never show the external exhaust trunking for a VIIC. This is due to a previous lack of understanding of snorkel construction, and the fact that their renderings were based on a fictional Type "VIIC 1944" design plan. The "VIIC 1944" design was based on a composite drawing by author Fritz Köhl (possibly in the late 1960s) that was intended to represent a VIIC with all possible snorkel modifications introduced before war's end. Unfortunately, this rendering was entirely fictional with no basis on actual snorkel construction diagrams or, to the best of this author's research, any planned modifications of the VIIC. This fictional rendering has been accepted as an actual wartime design and may have been the source of the incorrect restoration of U-995 (VIIC/41) in Laboe, Germany. U-995 was the last U-boat to be retrofitted with a snorkel and while it served in the postwar Norwegian Navy its external snorkel exhaust trunking was absolutely present. Period pictures exist in Eckard Wetzel's U995 (Motorbuch Verlag: Stuttgart, 2004), p.144–47. When the Norwegian government handed this U-boat back to the Federal Republic of Germany it was decided to remove the external starboard side exhaust trunking for some unknown reason and portray an inaccurate representation of a snorkel-equipped Type VIIC. Between the "Type VIIC 1944" plan and the incorrect restoration of U-995, this all-important external snorkel trunking shut-off valve on the starboard side of all VIICs has been ignored, forgotten, or dismissed by maritime archeologist, avocational historians, and model-makers alike.
10. Farago headed OP-16-Z within the Tenth Fleet. The US Tenth Fleet stood up in May 1943 as the first antisubmarine command in the US Navy's history with the express purpose to find, fix, and destroy German U-boats. OP-20-G, the office that produced the top secret Ultra intercepts, fell under the US Tenth Fleet command.
11. Ladislas Farago, The Tenth Fleet, The Story of the Submarine and Survival (Ivan Obolensky, Inc.: New York, 1962), pp.282-84. Farago makes it clear that due to a lack of intelligence and initial reports from German U-boat prisoners, the US Navy's Tenth Fleet believed the snorkel would not prove effective. Their assessment of the snorkel's capability quickly changed by the fall of 1944 as U-boat successes against Allied shipping mounted and their losses dropped to the lowest point in 18 months.
12. National Archives and Records Administration, Record Group 457, SRMN 037 U-boat Intelligence Summaries 1943–May 1945, March 16, 1945.
13. Sound Absorption and Sound Absorbers in Water (Dynamic Properties of Rubber and Rubberlike Substances in the Acoustic Frequency Region) by Walter Kuhl, Erwin Meyer, Hermann Oberst, Eugen Skudrzyk, and Konrad Tamm, collected by Erwin Meyer and translated by Charles E. Morgan, Jr. (Bureau of Ships: Washington, DC, June 1947). PRO/ADM/213-868 "Submarine Acoustic Camouflage.

Summary of Meyer and Oberst's account of the Laboratory Development and the Full Scale trials of the German project 'Alberich'" (Department of Physical Research, Admiralty) June 1948.

14 Arthur O. Bauer "Some hardly known aspects of the GHG, the U-Boat's group listening apparatus." pp.15–16.

15 Günter Hessler, The U-Boat War in the Atlantic 1939–1945 (Ministry of Defence, HMSO: London, 1992), Section 48-49, p.27. This work was written by Dönitz's son-in-law who served as his Staff Officer in the Operations section of the BdU staff during the war. After the war, while in captivity, he was granted access to a large amount of captured German documents, though not everything that might have been available, and wrote this history that amounts to the official German version of the U-boat war. While dated and incomplete, it still remains one of the best sources of German operational and tactical U-boat practices during the war.

16 Bundesarchiv-Militärarchiv/Wo4-13555.

17 Wesch rubber antiradar coating began to be applied to snorkel head valves and upper snorkel masts in the fall of 1944. Unlike Alberich, which was designed to counter sonar, Wesch matting was designed to reduce the radar signature of the exposed snorkel mast above water, making it harder for radar-equipped Allied vessels and aircraft to locate a snorkeling U-boat.

18 Royal Navy Archives, Gossport/1.A1990.197 "Relating to the 'Trials to compare Rubber-coated and non Rubber-coated Type VIIC U-Boats U-1105 and U-1171 respectively". Hereafter cited as RN Gossport/1.A1990.197. This file contains information on a variety of post-war U-Boat tests beyond that mentioned in the title.

19 Public Records Office, GCCS Naval Section: Decrypts hereafter cited as PRO/GCCS Decrypts/HW 18-401, 1124/13/6/44 and 1331/14/44.

20 Ibid.
21 Ibid.
22 Ibid.
23 Ibid.
24 Ibid.
25 Ibid.
26 Ibid.
27 National Archives and Records Administration, Record Group 38 Ultra Intercepts, Box 158, U-1105, hereafter cited as NARA/RG38/Box 158/U-1105.
28 PRO/GCCS Decrypts/HW 18-401.
29 Ibid.
30 NARA/RG38/Box 158/U-1105.
31 Ibid.
32 Ibid.
33 PRO/ADM 217-795, HMS CONN-Report of Proceedings 21st April-23rd May 1945.
34 Ibid.

35 Maryland Historical Trust, "The U-1105 Survey. A Report on the 1993 Archaeological Survey of 18ST636. A Second World War German Submarine in the Potomac River, Maryland" by Michael Pohuski and Donald Shomette, December 1994, p.15, hereafter cited as MHT/1993 U-1105 Survey.

36 Henry Keatts, Dive into History: U-Boats (Pisces Books: Houston Texas, 1994), p.176.

37 Report Loch No. 7/45/15 Group. Surrender of U-Boats 1945. Author's collection.

38 NARA/RG242/T1022/R3673/U-1305 KTB.

39 Derek Waller, "U-1105 in the US Navy – 1945 to 1949", p.1. Unpublished paper on the history of U-1105.

40 Royal Navy Archives, Gosport/1.A1990.197 "Relating to the 'Trials to compare Rubber-coated and non Rubber-coated Type VIIC U-Boats U-1105 and U-1171 respectively". Hereafter cited as RN Gosport/1.A1990.197. This files contains information on a variety of postwar U-boat tests beyond that mentioned in the title; hereafter cited as RN Gosport/1.A1990.197.

41 Ibid.
42 Waller, p.3.
43 Ibid.
44 PRO/ADM 173-20075, Monthly Log of U-1105, August 1945.
45 Waller, p.3.
46 PRO/ADM 116-5571b.
47 PRO/ADM 173-20075.
48 Ibid.
49 PRO/ADM 116-5571a.
50 PRO/ADM 173-20076, Monthly Log of U-1105, September 1945.
51 PRO/ADM 116-5571b
52 Pro/ADM 173-20077, Monthly Log of U-1105, October 1945.
53 PRO/ADM 116-5567, U-1105.
54 Ibid.
55 Ibid.
56 PRO/ADM 116-5571, TNC UK Inspection Admiral Archer Report, September 25, 1945.
57 Waller, p.5.
58 PRO/ADM 173-20077.
59 PRO/ADM 116-5567.
60 PRO/ADM 173-20077.
61 PRO/ADM 116-5567.
62 Ibid.
63 PRO/ADM 204-1610, Detection of German Schnorchel Exhausts by Thermal Radiation, October 11, 1945.
64 PRO/ADM 204-1289, Noise Trials of German U-Boats. Types VIIC & XXIII, May 25, 1946.
65 PRO/ADM 259-434, Comparison of V.F.E. (Variable Field Effect) Signals from Rubber-Covered and Non-Rubber-Covered U-Boats, December 1, 1945.

66 RN Gosport/A1986.010, "The Latter Time in My Submarine Service 1944-1946" by J. E. Woodcock.
67 The application states: "Not only does this vessel represent the only known example of this early stealth technology but its study directly provided the data that led to subsequent treatment of American and Soviet submarines." U-1105 Site File (18ST636), Maryland Historical Trust, United States Department of the Interior National Park Service, National Register of Historic Places, Registration Form.
68 NARA US Navy Technical Mission in Europe Report No. 352-45, Rubber Covering of German Submarines Anti-Asdic (German Code Name "Alberich") dated September 20, 1945, Appendix; hereafter cited ad NARA/NTME/352-45.
69 Video transcript provided by the Murphy family.
70 Capt Hubert Murphy's efficiency report for the period dated March–May 1945.
71 Maryland Historical Trust Records. Letter from William Ferguson dated December 26, 1985.
72 Admiralty message dated 10.12.45 from C.inC. Plymouth to C.inC. Portsmouth. PRO. [U-1105 file]
73 PRO/ADM 259-434, Comparison of V.F.E. (Variable Field Effect) Signals from Rubber-Covered and Non-Rubber-Covered U-Boats, December 1, 1945. RN Gosport/A1986.010, "The Latter Time in My Submarine Service 1944-1946" by J. E. Woodcock.
74 3/19/46 Officer Fitness Report. Lt Cdr Murphy provided by the Murphy family.
75 National Archived Records Administration/RG 24, USS Hoist, USS Salvager, and USS Windlass Deck Logs. Confidential letter to Commander, US Navy Shipyard, Portsmouth N.H. Ex-German Submarines Preservation preparatory to Underwater Explosion Tests, Request for Comments regarding.
76 MHT/1993 U-1105 Survey.
77 NARA/RG 24/USS Hoist, USS Salvager and USS Windlass Deck Logs. Letter from CNO to CINCLANT dated February 10, 1946.
78 Robert Gannon, Hellions of the Deep: The Development of American Torpedoes in World War II (Penn State University Press: New York, 2009), p.58.
79 http://asa.aip.org/encomia/gold/bolt.html
80 MHT/1993 U-1105 Survey.
81 NARA/RG 24/USS Hoist, USS Salvager and USS Windlass Deck Logs.
82 Ibid.
83 Ibid.
84 Ibid.
85 NARA RG38 CNO Command Files Box 528/U-1105.
86 MHT OP33P2/Sa Navy Department memo from Chief of Naval Operations to Commander in Chief, US Atlantic Fleet. Test Depth Charge and Salvage Equipment on U-1105. Dated August 2 1948. MHT Document.
87 NARA/RG24/USS Hoist Deck Log, August 1948.
88 MHT/1993 U-1105 Survey.
89 NARA/RG24/Salvager Deck Log.
90 NARA/RG24/Windlass Log.
91 Washington Post, October 13, 1948. P.2.
92 NARA/RG24/Salvager Deck Log.
93 NARA/RG24/Windlass Deck Log.
94 MHT/1993 U-1105 Survey. Memorandum. Commander Task Unit 49.4.3 to Commander Service Force, US Atlantic Fleet. August 4, 1949. Salvage of Ex-German Submarine U-1105, Report of Progress, Period July 11 to August 4, 1949.
95 MHT/1993 U-1105 Survey. Memorandum. Commander Task Unit 49.4.3 to Commander Service Force, US Atlantic Fleet. August 4, 1949. Salvage of Ex-German Submarine U-1105, Report of Progress, Period July 11 to August 4, 1949.
96 MHT/1993 U-1105 Survey. Memorandum to Commander Service Force, US Atlantic Fleet. Salvage Operations Report of Progress. August 20, 1949.
97 Film, "Close is Near Enough", Naval Research Development Center and the Naval Ordnance Lab (1973).
98 MHT/1993 U-1105 Survey and Explosive and Structural Test of U-1105 German Submarine, Operational Report, October 19, 1949.
99 MHT/1993 U-1105 Survey.
100 Interview conducted by the author with a long-time resident and diver of the Chesapeake Bay. Name being withheld upon request.
101 Several attempts were made to contact Uwe Lovas for an interview without success. He has reportedly informed others that he "is done with U-1105."
102 NAVPERS 1616-B, The Submarine (ComSubLant, Standards and Curriculum Division, Training, Bureau of Naval Personnel: 1961), p.183. Reprint edition. (PerriscioeFilm.com: 2008)
103 In May 1947 the US Navy conducted its first snorkel endurance cruise with the USS Irex (SS-482). During that cruise, the Irex achieved a transatlantic snorkel cruise of 20 days underwater. This was a widely celebrated success for the US Navy. It came nearly four years after German U-boats set underwater endurance records three times that length. As examples, U-826 cruised for 64 days submerged and U-293 for 83 days in 1944–45. It took the US Navy until the 1950s before US snorkel-equipped submarines routinely cruised submerged for a month or more.

INDEX

acoustic camourflage *see* Alberich rubber coating; anechoic tiles
Acoustics Laboratory, MIT 92
Admiralty 62, 69, 73, 77
Air Defense School VII training, German 40–1
Alberich rubber coating 14, 15, 28–31, *29, 30,* 50, 51, 56–8, *61,* 62, 63–4, 67, 69, 70, 73–6, *74, 75,* 76, 78, 79, 80–2, 85, 89, 90–1, 104, 110–11, *118*
Allied forces 15, 16, 17–18, 26–7, 28, 29, 31, 33, 34, 36, 37, 47, 62–3
see also Royal Navy; Soviet Union; United States Navy
Allied Submarine Detection Investigation Committee (ASDIC) devices 28, 29, 31, 47, 51, 54–5, 78, 111
ammunition container, *U-1105*'s *115, 121*
anechoic tiles 77, *77,* 83
antisubmarine trials, postwar 63–4
archaeological surveys of *U-1105* 24, 103–5, 109–25, *111–25*
  Alberich coating 110–11
  ammunition container *115, 121*
  conning tower 109, 110–11, *111,* 112–16, *112–16,* 118–20, *118–20*
  hatches 125, *125*
  light features *117, 118, 122, 123*
  periscopes *114, 116, 117, 122, 124*
  snorkel exhaust trunking 14, 24, 65, *65, 68,* 105, 119, *119,* 120, *128*
  wave guard *114*
  Wintergarten and associated features 109, 121, *121,* 122, *122,* 123, *123*
  wooden features *116, 121*
attack periscope, *U-1105*'s *59, 116, 117, 122, 124*

Bay of Biscay 18
Befehlshaber de U-Boote (BdU) 16, 26, 27, 29, 40, 41, 47, 48, 105
'Black May' (1943) 17–18, 35
Bolt, Dr H. 90, 91–2
bottoming tactics 26–8, 57, 59–60
Bureau of Ships, US Navy 80–1, 82, 89, 90, 92, 93–4, 95
Byron, HMS 53, 54

carbon monoxide exposure 22–4
Coll Sanctuary tests 69
Conn, HMS 50, 53, 54–6, 60
conning tower, *U-1105*'s 14, 30, 31, 63, *63, 65, 65,* 66, *68,* 69, 84, *84,* 86, 87, 97, *97,* 106, 108, 109, 110–11, *111,* 112–16, *112–16,* 118–20, *118–20*
convoys, Allied 16, 18, 47, 54

Deadlight, Operation 105
Deane, HMS 53, 54
depth charges/testing 27, 34, 51, 52, 54, 56, 57, 89, 95–6, *97, 98,* 99–101, *100, 101*
depth-keeping gear, U-boat 36
diving the *U-1105* today 12–14, 106–8
  see also archaeological surveys of *U-1105*
Donitz, Adm Karl 17–18, 35–6, 58, 105

Earner, HMS 84
Electro-boats 63
Enigma code 17
  see also Ultra, Operation

Farago, Ladislas 26
Ferguson, William 84–5, 87, 88–9
Fitzroy, HMS 53, 54
flash transmission systems 36–7
FuMO 61 Hohentwiel radar access 36, *115*

Germaniawerft engines 23
GHG Balkon 15, 32–3, *33,* 35, 36, 50, 57, 59, *75,* 85, 104
Ghormley, Adm Robert 73–4
Greater Underwater Propulsion Power Program (GUPPY) 110
Greenling, USS 83, 84
guns, *U-1105*'s 35, 36

hatches, *U-1105*'s 86, 87, 125, *125*
Hedgehog rockets 27, 56, 57
Hessler, Günter 34–5
'High Tea'/sonobuoys 53–4
Hoist, USS 95
Holy Loch 61, *61,* 63, *63,* 65, *65, 68,* 69, 70, 71, 72
Holyhead 72–3
'Hunter-Killer' task forces, US 26, 54

Kingfisher, HMS 69
Kurier flash transmission systems 36–7

Levchenko, Adm Gordei 74
light fittings, *U-1105*'s *117, 118, 122, 123*
Lisahally, Londonderry 61, 62, 69–70
Loch Alsh 60, 61
Loch Eriboll 52, 53, 59, *59,* 60
Loch Goil 64, 66, 67, *67,* 70, *70,* 71–2, *72*
Loch Ryan 62, 75–6
Lovas, Uwe 102

Malorny, Dr Guenther 22, 23–4
MAN diesel engines 23
Massachusetts Institute of Technology (MIT) 82, 90, 91–2
microphones, underwater 91
Monmouth Coast, SS 53
Murphy, LCdr Hubert T. *83,* 83–6, 89, 125

Naval Research Lab (NRL), US 67, 81–2, 90–1
NAVPERS 16160-B The Submarine 110
Nelson Salvage and Construction Company 102
Nordseewerke Shipyard, Emden 34

passive sonar arrays 15, 32–3, *33,* 35, 36, 50, 57, 59, *75,* 85, 104
Paukenschlag, Operation 91
Piney Point, Chesapeake Bay 8, 13, 95–101, 102
Pohuski, Michael 103–4
Potomac River *see* Piney Point, Chesapeake Bay

radar/antiradar 17, 18, 27–8, 36, 63, 64, 69, 90, *115*
radio systems 36–7
Redmill, HMS 50, 52, 53, 54–7, *56*
Rettungsbehalter (rescue container) 122
Royal Air Force 53–4, 58–9
Royal Navy 15, 37, 50–1, 52–61, *55,* 91
  evaluation and testing *U-ll05* 62–90
Rupert, HMS 53, 55–6, 61

salvage testing 95–6, 119

Salvager, USS 95, 96
Schade, Cdr Henry A. 82
Schwarz, Klt Hans-Joachim 37–45, *39, 40, 46–50,* 51–2, 55, 57, 59, 60–1
Shiffsbuches documents 35
Shomette, Donald 103–4
sky periscope, *U-1105*'s *114, 116*
Snapper, USS 83, 84
snorkel-cams 23–4
snorkels 15, *17,* 18, *19, 20,* 20–6, *24,* 25, 27–9, 30, 32, 33, 36, 37, 40, 41, 45, 48, 57, 59–60, 62, 63, 64, 67, *67,* 69, 70, 71–2, 78, 79, 86–7, 97, 104–5, 110
   exhaust trunking 14, 22–4, *24,* 65, *65,* 68, 69, 105, 110, 119, *119, 128*
sonar/antisonar 15, 26–7, 28, 31, 32–3, 36, 64, 67, 69, 78, 85, 104
   see also Alberich rubber coating
sonobuoys (code word 'High Tea') 53–4
Soviet Union 73–7, 82, 85, 105, 111

'Total Undersea War' 15, 18, 105
training, German U-boat 40–1, 42–4
Tripartite Naval Commission (TNC) 72, 73, 74–6
Tuna, HMS 84
21st Escort Group, Royal Navy 37, 50–1, 52–8, *55,* 61, 111

U-485 67, 70, 75, 80, 91
U-505 92, 105
U-671 24
U-826 63, *63*
U-853 100
U-901 46–7
U-1010 46–7
U-1105 'Black Panther' (Type VIIC) 25, *39, 40, 84, 104, 105*
   Alberich coating 14, 15, 29, *29, 30,* 31, 36, 50, 51, 56–8, *61,* 62, 63–4, 67, 70, 73–4, *74, 75,* 76, 78, 79, 80, 85, 89, 90, 104, 110–11, *118*
   ammunition container *115, 121*
   archaeological surveys 24, 103–5, 109–25, *111–25*
   combat patrol and engagement 46–51, *49,* 53–8
   conning tower 14, 30, 31, 63, *63,* 65, *65, 66, 68,* 69, 84, *84,* 86, 87, 97, *97,* 106, 108, 109, 110–11, *111,* 112–16, *112–16,* 118–20, *118–20*
   conning tower emblem 66

construction of 34, 35–6
demolition scheduled 93–4
depth charge testing in the US 95–6, *97, 98,* 99–101, *100, 101*
diving the wreck 12–14, 106–8
end of the war and surrender of 25, 51–2, 58, 58–61, *61, 63*
flash transmission system 36–7
German captain's reports 37–45, 46–50, 51–2, 57, 59, 60–1
German crew training 40–1, 42–4, 45
GHG Balkon 15, 33, *33,* 35, 36, 50, 57, *75,* 104
guns 35, 36, 121, *121, 122*
light fittings *117, 118, 122, 123*
misidentification of type 34–5
name plate *124*
Operation *Ultra* intercepts 40, 41–4, 45, 48
periscopes *114, 116, 117, 122, 124*
post-war testing in the US 90–101, *93, 97, 98, 100, 101,* 104
postwar evaluation and testing in the UK 62–4, 65, *65,* 66, *66, 67, 68,* 69–74, *70, 74, 75,* 76–9
radar and radar detectors 25, 36, *115*
recorded as a wreck 102–3, *103*
RN antisubmarine trials 63–4, 65, *65, 66, 66, 67, 68,* 69
RN crewman's recollections of testing 78–9
RN Pennant Number 69
salvage testing in the US 95–6, 119
Shiffsbuches documents 35
snorkel 15, 36, 37, 40, 41, 45, 48, 59–60, 62, 63, 65, *65,* 69, 70, 71–2, 78, 79, 86–7, 97, 104–5, 110, 120, *120*
   exhaust trunking 14, 24, 65, *65, 68,* 69, 105, 119, *119,* 120, *128*
Soviet and US interest in 73–4, 76, 82, 85, 105
transatlantic crossing 76, 77, 83, 86–90, 105, 125
unique watertight storage tube 65, *65,* 84, *94, 94,* 123, *123, 128*
Wintergarten and associated features 35, 65, *65, 68,* 69, 72, *72,* 84, *84,* 94, *94,* 109, 121, *121,* 122, *122,* 123, *123*
wooden features *116, 121*
U-1171 'White Puma' (VIIC) 63, 67, *67, 68,* 69, 70, 73, 76, 77, 78
U-1228 46
U-1305 53–4, 57, *59,* 59–60
U-2326 63, 78

U-2502 63
U-2513 93
U-boats 14–15, 63–4
   bottoming tactics 26–8, 57, 59–60
   carbon monoxide build-up 22–4
   depth-keeping gear 36
   flash transmission systems 36–7
   operations 25
   resilience to depth charges 100
   weaknesses 16–17, 22–4
   wolfpack tactics 16, 26
   see also Alberich coating; GHG Balkon; snorkels; individual U-boats by name
Ultra, Operation 17, 37, 40, 41–4, 45, 48, 54
Underwater Electric Potential (UEP) Fields study 95
Underwater Explosive Research Division 99
United Nations 62
United States Navy 12, 26, 35, 73–4, 76, 77, 110
   Alberich/acoustic camouflage 15, 67, 73–4, 76, 80–2, 91–2, 111–12
   Bureau of Ships 80–1, 82, 89, 90, 92, 93–4, 95
   depth charge testing 89, 95–6, *97,* 98–100, *100*
   Greater Underwater Propulsion Power Progam (GUPPY) 110
   Naval Research Laboratory (NRL) 67, 81–2, 90
   Ordnance Disposal Unit and School 102–3
   salvage testing 95–6, 119
   U-1105's transatlantic crossing 76, 77, 83, 86–90, 105, 125
   Underwater Explosive Research Division 99

Walter, Dr Hellmuth 18
Wanze 2 radar detectors 36
Washington Post 96
wave guard, *U1105*'s conning tower *114*
Wesch antiradar matting 25, 36, 63, 74–5
Wilks, ERA Roy 86, 88
Windlass, USS 91, *91,* 95, 96
Wintergarten (Turmumbau II) and associated features, *U-1105*'s 35, 65, *65, 68,* 69, 72, *72,* 84, *84,* 94, *94,* 109, 121, *121,* 122, *122,* 123, *123*
wolfpack tactics 16, 26
Woodcock, 3rd Class ERA John Edge 78–9, 86–8